CARBON REMOVAL

HOWARD J. HERZOG AND
NIALL MAC DOWELL

The MIT Press | Cambridge, Massachusetts | London, England

The MIT Press
Massachusetts Institute of Technology
77 Massachusetts Avenue, Cambridge, MA 02139
mitpress.mit.edu

The MIT Press would like to thank the anonymous peer reviewers who provided comments on drafts of this book. The generous work of academic experts is essential for establishing the authority and quality of our publications. We acknowledge with gratitude the contributions of these otherwise uncredited readers.

This book was set in Chaparral Pro by New Best-set Typesetters Ltd. Printed and bound in the United States of America.

Library of Congress Cataloging-in-Publication Data is available.

ISBN: 978-0-262-55136-6

10   9   8   7   6   5   4   3   2   1

EU Authorised Representative: Easy Access System Europe, Mustamäe tee 50, 10621 Tallinn, Estonia | Email: gpsr.requests@easproject.com

CARBON REMOVAL

The MIT Press Essential Knowledge Series

A complete list of books in this series can be found online at
https://mitpress.mit.edu/books/series/mit-press-essential-knowledge-series.

CONTENTS

SERIES FOREWORD

The MIT Press Essential Knowledge series offers accessible, concise, beautifully produced pocket-size books on topics of current interest. Written by leading thinkers, the books in this series deliver expert overviews of subjects that range from the cultural and the historical to the scientific and the technical.

In today's era of instant information gratification, we have ready access to opinions, rationalizations, and superficial descriptions. Much harder to come by is the foundational knowledge that informs a principled understanding of the world. Essential Knowledge books fill that need. Synthesizing specialized subject matter for nonspecialists and engaging critical topics through fundamentals, each of these compact volumes offers readers a point of access to complex ideas.

In 2018, the MIT Press published a book in its Essential Knowledge series entitled *Carbon Capture*. That book discussed technologies that captured carbon dioxide (CO_2) produced by power plants and industrial facilities *before* it was emitted into the atmosphere. This book on *Carbon Removal* explores pathways to remove CO_2 *after* it has already been emitted into the atmosphere. Carbon capture and carbon removal are two distinct approaches to mitigating climate change, with each approach having a multitude of technological options. Some of the technologies used for carbon removal overlap with some of those used for carbon capture, which causes confusion at times because people may use the terms *capture* and *removal* interchangeably. However, in many ways, the world of carbon removal that we explore in this book is very different from the world of carbon capture. For example, carbon removal encompasses a wide range of approaches, including removal by both biological and chemical pathways, as well as carbon storage in vegetation, soils, the ocean, and geological formations.

To understand the current interest in carbon dioxide removal (CDR), we must travel back in time to June 1992, when representatives from 179 countries came together in Rio de Janeiro, Brazil, at what is known as the Earth Summit. The meeting's primary objective "was to produce a broad agenda and a new blueprint for international action on environmental and development issues that would help guide international cooperation and development policy in the twenty-first century."[1] There were several important outcomes of the meeting, including the drafting of the United Nations Framework Convention on Climate Change (UNFCCC). Article 2 of the convention defines its objective: "to achieve . . . stabilization of greenhouse gas concentrations in the atmosphere at a level that would prevent dangerous anthropogenic interference with the climate system."[2]

The UNFCCC was signed by President George H. W. Bush and ratified by the United States Senate. Today, 197 countries are Parties to the Convention. These countries gather once a year at a Conference of the Parties, commonly called a COP. The UNFCCC was deliberately vague on many important issues, leaving it up to the COPs to develop rules and protocols to carry out the stated objective. For technical input to the process, the COP relies heavily on reports generated by the Intergovernmental Panel on Climate Change (IPCC).[3]

One open question was how to define and quantify the stabilization level that "would prevent dangerous anthropogenic interference with the climate system." Through the COP and IPCC processes, in 2015 the international community arrived at a consensus that greenhouse gases in the atmosphere must be kept at such a level to keep global temperatures from rising more than 1.5°C (degrees Celsius) to 2°C compared to pre-industrial times. Unfortunately, it is becoming abundantly clear that we are going to exceed that level. We are likely to reach 1.5°C by 2030, and, unless significantly more stringent climate policies are implemented around the world, we will almost certainly exceed the 2°C target.

Carbon dioxide is the greenhouse gas that contributes most to the warming. People are realizing that since we are putting too much CO_2 into the atmosphere, the only way to eventually meet the stabilization target is to remove CO_2 from the atmosphere. This has led to a frenzy of development of CDR technologies. For example, on June 8, 2022, a *Wall Street Journal* article headlined "Carbon-Removal Industry Draws Billions to Fight Climate Change" stated "In the past few months, tech giants like Google and Facebook, consulting firms McKinsey and BCG, financial firms UBS and Swiss Re, plus the royal family of Liechtenstein, have promised to pay generously for carbon that is removed from the atmosphere and stored."[4]

While some CDR options like afforestation are relatively inexpensive, at approximately $20 per metric tonne of CO_2 (tCO_2), others (like direct air capture) are very expensive today (hundreds of dollars per tCO_2 or more). It will be important to distinguish between the CDR service provided by afforestation, direct air capture, and other CDR options, as it is not obvious that they are perfectly fungible with fossil carbon emissions.

One company, Climeworks, will directly remove CO_2 from the atmosphere for you for a price of $1,500 per tonne of CO_2.[5] As a comparison, one of the largest markets for keeping CO_2 out of the atmosphere is the European Union (EU) Emissions Trading System (ETS), where the average market price in 2023 was about $100 per tonne of CO_2, a factor of 15 less than Climeworks' selling price.[6] Yet according to the *Wall Street Journal* article, Climeworks has privately raised $650 million. Are these investors smarter than the rest of us, or is this just another financial bubble waiting to burst?

Today the world of CDR is part gold rush, part wild west. There are few rules for companies operating in this field. For example, CDR requires energy, which, depending on its source, may have CO_2 emissions associated with it. Any emissions generated during CDR must be subtracted from the gross amount captured. In other words, you should be allowed to take credit for only the net amount of CO_2 removed, not the gross amount. However, there are no official accounting rules for CDR today. Companies can claim whatever they want.

The goal of this book is to explore the new and rapidly developing world of CDR. We describe the science behind the various CDR approaches, explore how the leading CDR technologies work, and look at the politics involved in making CDR a reality. It is clear that CDR has the potential to play an important role in meeting our climate change goals. However, it is also clear that many challenges must be addressed to move CDR from today's speculation to deployment at scale. In this book, we aim to provide an objective analysis of the opportunities and challenges for CDR and to separate myth from reality.

CARBON DIOXIDE REMOVAL

Over the years, humanity has become profoundly good at extracting fossil fuels from the earth. In 2021, total global energy consumption was 595 exajoules (EJ).[1] Of this, 82 percent was provided by fossil fuels—coal, oil, and natural gas.[2] This led to the emission of over 37 billion tonnes (gigatonnes or Gt) of carbon dioxide (CO_2)—an increase of approximately 65 percent since the 1992 Earth Summit.[3] In addition, significant quantities of CO_2 are emitted from nonenergy sources, such as cement production, land use, and land-use change. Moreover, other greenhouse gases (GHGs), including methane (CH_4) and nitrous oxide (N_2O), also contribute to climate change. Adding all GHG emissions together, with each gas weighted by its impact on climate change relative to CO_2, yields the total GHG emissions in terms of CO_2 equivalent. In 2022, total anthropogenic GHG emissions exceeded 53 Gt of CO_2 equivalent,[4] of which 69 percent was CO_2.

In order to limit global mean temperature rise, many countries have pledged to reduce their total global anthropogenic greenhouse gas emissions to net zero by between 2050 and 2070. In this context, net zero means that total GHG emissions need to be balanced by an equivalent amount simultaneously removed from the atmosphere. It is important to recognize that reducing total global GHG emissions to net zero is implicit in stabilizing the climate *at any level* of temperature rise. All that varies is the CO_2 atmospheric concentration at which we choose to do this.

As temperature rise is a function of atmospheric CO_2 concentration, this leads to the concept of a *carbon budget*. This is the total cumulative amount of CO_2 emissions consistent with a given temperature rise. To put this in some perspective, humanity has emitted over 2,400 gigatonnes of CO_2 since the pre-industrial era. While there are various estimates, each with their own level of confidence, as of the beginning of 2022, the remaining carbon budget for 1.5°C was 420 gigatonnes of CO_2, or eleven years of emissions if they were held constant at 2021 levels. The carbon budget for 2°C warming was 1,270 gigatonnes of CO_2.[5]

In terms of meeting climate targets, directly mitigating (i.e., reducing) emissions is the primary focus. Mitigation has historically involved improving efficiency, switching to less carbon-intensive fuels, using nuclear power, deploying renewable energies such as solar and wind, and installing carbon capture and storage (CCS). However, it will not be cost-effective to directly mitigate 100 percent of GHG emissions. Some are too small, too remote, or simply too costly to mitigate with existing technology. With others, it may be possible to mitigate almost all, but eliminating the final few percent may be very challenging and costly.

Similarly, it is not obvious that humanity will stay within the remaining carbon budget. At the time of this writing in February 2025, the absolute amount of fossil fuel used in the world's energy systems is still increasing.

Thus, CDR can be considered to perform two services in helping us meet long-term climate targets—compensating for hard-to-abate emissions and compensating for any overshoot of the carbon budget. In this context, the overall problem can be thought of using the "bathtub" analogy, as illustrated in figure 1.

While there are a number of GHGs in the atmosphere, CO_2 is present in the highest concentration and is by far the easiest to remove. Methane, for example, is present in the atmosphere but only at the parts per billion (ppb) level. This is about two hundred times less abundant than CO_2, which is at the parts per million level (ppm). Hence, efforts to reduce the concentration of GHGs in the atmosphere predominantly focus on CO_2,[6] leading to the term *carbon dioxide removal*, or *CDR*. All of this taken together lets us conclude with a definition of CDR—namely, "the physical removal of CO_2 from the atmosphere in a manner intended to be permanent."[7]

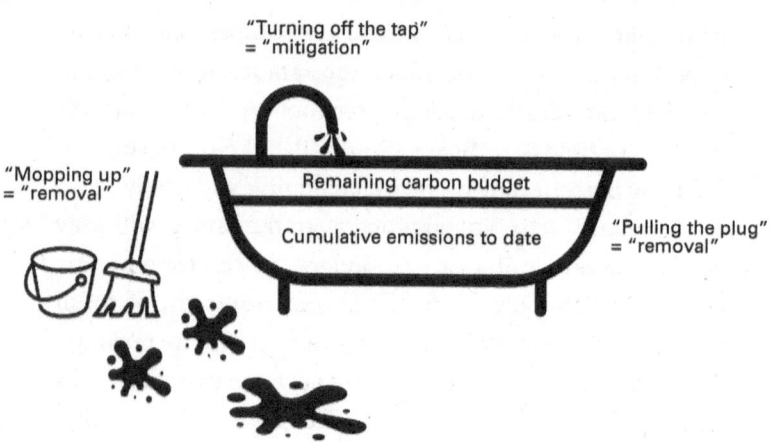

Figure 1 In this analogy, the atmosphere can be thought of as a bathtub, with the tap (representing anthropogenic processes) putting water (greenhouse gases) into the bathtub (the atmosphere). In order to avoid overflowing the tub, we can both turn off the tap (mitigate emissions) and also pull the plug (carbon dioxide removal). The bathtub may yet overflow, which corresponds to exceeding the carbon budget for a given climate goal, so additional CDR may be required to "mop up" potential spillage.

A Brief History of Carbon Dioxide Removal

As corporate and national commitments to net-zero targets proliferate, the concept of removing carbon dioxide from the atmosphere has been gaining increasing attention from many stakeholders, including industrial, political, financial, and academia. However, it is a much more established concept than we might think. The concepts of bioenergy with carbon capture and storage (BECCS) and

CO_2 disposal in carbonate minerals (the precursor to what is today referred to as *enhanced rock weathering*) were both actively discussed from the mid-1990s. Direct air capture (DAC) was proposed from the late 1990s. Afforestation and reforestation projects were part of the Clean Development Mechanism in the 1997 Kyoto Protocol.

This paradigm shifted rapidly in the 2000s. In 2008, the United Kingdom committed to an 80 percent reduction in economy-wide emissions by 2050 compared to 1990 levels—an unprecedented commitment at that time. However, even then the need for "at-scale" deployment of CDR was unclear. Much of the required mitigation could be achieved via the relatively conventional approaches of energy efficiency, fuel switching, increased use of nonfossil energy sources, and carbon capture and storage. At that time, CDR—mostly in the form of BECCS—had made its way into various *integrated assessment models* (IAMs) and was not, initially, overly scrutinized. However, as time went on, observations regarding the viability of various model outcomes became more common, leading to increased emphasis on additional pathways to the removal of greenhouse gases from the atmosphere.

It was not until the 2015 Paris Agreement and the subsequent special report by the IPCC on 1.5°C in 2018 that the need for the at-scale deployment of CDR became clear. To put *at scale* in context, contemporary total anthropogenic GHG emissions are on the order of 50 gigatonnes of

CO_2 equivalent per year. By 2100, many integrated assessment models forecast that on the order of 10 $GtCO_2$ to 20 $GtCO_2$ equivalent per year of CDR will be required to meet our stabilization goals. This means we will need to develop a CO_2 removal industry that is comparable in scale to the contemporary fossil fuel industry.

Since about 2019, national and corporate pledges to reduce total GHG emissions to net zero by between about 2040 and 2070 have abounded. Where pathways to achieve these goals have been articulated, there has often been substantial reliance on CDR pathways to compensate for hard-to-abate sections of the economy or for elements of a corporate supply chain over which the corporation with the net-zero pledge has limited or no agency. This increased focus on CDR has also led to significant levels of concern that CDR is, at best, a dangerous distraction or, at worst, a cynical effort to "greenwash" operations and essentially enable business as usual to continue.[8] Some of this concern appears rooted in an essentially philosophical interpretation of climate goals. Is the aim to reduce net CO_2 to zero and remain agnostic about technologies or fuels, or is the aim to "defossilize" the global economy by expeditiously switching entirely to nonfossil energy sources, primarily wind, solar, and hydro? Choosing between these goals largely comes down to national policy and personal predilection.

One justifiable reason for concerns about CDR is the dubious veracity of much of what has historically been

It was not until the 2015 Paris Agreement and the subsequent special report by the IPCC on 1.5°C in 2018 that the need for the at-scale deployment of CDR became clear.

sold in the marketplace. Among other issues, people have questioned the accounting methods used to calculate the reductions, as well as the durability of the removals. For CDR to play a meaningful role in meeting climate commitments, these concerns will need to be addressed via a combination of national and international policy and regulation and via technological interventions on the monitoring, reporting, and verification (MRV) of the various removal technologies and their associated carbon sinks.

An Overview of Carbon Dioxide Removal Pathways

While CDR pathways span a wide range of technologies and practices, they all must have two common characteristics. First, they must remove CO_2 from the atmosphere. Second, they must securely store the carbon either as CO_2 or in another form so it stays out of the atmosphere for a "significant" amount of time. Later in this book, we explore what qualifies as a significant amount of time.

CO_2 can be removed from air in one of two ways—biologically or chemically. The biological pathway lets trees and plants do the work through photosynthesis. During photosynthesis, plants convert solar energy into chemical energy in the form of sugars and starches. The three main ingredients required for this process are sunlight, water, and CO_2. The net effect is that the CO_2 is removed from the air and incorporated into the plants' biomass. Energy for this process is provided for free by the sun.

The chemical pathway uses chemicals (referred to as *sorbents*) to remove CO_2 from the air. Since CO_2 is a weak acid, it will be attracted by chemicals that are basic (i.e., alkaline) in nature. In this pathway, air flows over an alkaline sorbent, which will chemically react with the CO_2, removing it from the air. In general, these processes require significant amounts of energy.

Once removed from the air, CO_2 can be stored in one of four places—the terrestrial biosphere (i.e., vegetation and soils), the oceans, rocks, or deep geological formations. The terrestrial biosphere stores the carbon in the form of organic carbon, while in the ocean the CO_2 is primarily stored as inorganic carbon in the form of bicarbonate and carbonate ions dissolved in the seawater. To form rocks, CO_2 reacts with certain minerals, such as serpentine or olivine, and gets incorporated into the rock in the form of carbonates. Finally, CO_2 can be injected as a supercritical fluid (essentially a high-pressure liquid, >80 bar) into deep underground formations, similar to the formations that have trapped oil and gas for millions of years. These various storage options differ substantially in terms of their durability, measurement and verification techniques, and susceptibility to reversal, either by natural means (e.g., wildfires) or human intervention (e.g., deforestation). More details on these four storage options are presented in later chapters.

CO_2 can be removed from air . . . biologically or chemically. . . . Once removed from the air, CO_2 can be stored in one of four places—the terrestrial biosphere, the oceans, rocks, or deep geological formations.

Table 1   Carbon dioxide removal pathways

CO_2 removal mechanism	CO_2 storage medium	Example pathways
Biological	Terrestrial biosphere (vegetation and soils)	Afforestation and reforestation Modified agricultural practices Biochar
	Ocean	Iron fertilization
	Deep geologic formations	Bioenergy with carbon capture and storage (BECCS)
Chemical	Ocean	Ocean alkalinity enhancement
	Rocks	Enhanced rock weathering
	Deep geologic formations	Direct air capture (DAC)

Many CDR pathways have been proposed to remove the CO_2 from the air and store it away from the atmosphere. Among other characteristics, these pathways vary significantly in their complexity, scale, and costs. Table 1 lists some of the better-known CDR pathways.

All of these pathways and more are discussed in the subsequent chapters. Chapter 2 looks at the natural carbon cycle, where nature moves carbon between the atmosphere and the four storage reservoirs listed above. Before we look at specific CDR pathways, we first explore the role CDR can play in a climate change portfolio in chapter 3. Then, chapter 4 looks at what is referred to as enhancing natural sinks. This means increasing the amount of carbon stored in the terrestrial biosphere and the oceans.

Chapter 5 discusses biomass-based carbon removal and storage including BECCS and biochar. Chapter 6 explores engineered removal pathways, which includes direct air capture and enhanced rock weathering. Ocean-based removal pathways, including ocean alkalinity enhancement and iron fertilization, are described in chapter 7.

Several important issues impact the effectiveness and feasibility of the various CDR pathways. One is to get the accounting correct by doing a life-cycle greenhouse gas balance. Another issue is additionality, meaning that any reductions made by a CDR project need to be "additional" compared to the case in which the project had not been implemented. The issue of permanence deals with how long a reservoir can keep CO_2 out of the atmosphere. These issues are highlighted in chapter 8, where we compare the various carbon removal pathways.

The book concludes by looking into the future and discussing what is needed for CDR to deploy at large scale. While there is much uncertainty, it is already clear that CDR can play an important role in getting us to net zero. In this book, we explore the various CDR pathways to better understand what we can realistically expect. We hope readers of this book come away with a good understanding of the fundamentals of CDR so they can make their own informed assessment of this rapidly changing field.

THE CARBON CYCLE

Understanding how carbon dioxide removal (CDR) works first requires understanding the carbon cycle, whereby carbon is continuously exchanged between the atmosphere, the ocean, and the terrestrial biosphere (i.e., vegetation and soils). The carbon cycle is best described in units of carbon (C), not carbon dioxide (CO_2). This is because the carbon is stored in many different forms—primarily carbon dioxide in the atmosphere, organic carbon (e.g., biomass) in the terrestrial biosphere, and carbonate ($CO_3^=$) and bicarbonate (HCO_3^-) ions in the ocean. One kilogram of carbon dioxide contains 0.273 kilograms of carbon.

Carbon Stocks and Flows

The bathtub analogy (see figure 1, chapter 1) is often used to help people visualize the carbon cycle (see figure 2). As figure 1 shows, the amount of carbon in the atmosphere is analogous to the amount of water in the tub. Just as water flows into the tub from the faucet or out of the tub through the drain, carbon flows into and out of the atmosphere, primarily through exchanges with the ocean and terrestrial biosphere. This analogy also demonstrates the concept of "stocks" (the amount of water in the bathtub) and "flows" (the amount of water coming from the faucet or leaving by the drain).[1] Only when the flow into the bathtub equals the flow out of the tub will the stock of water in the tub be stable. The same is true of the atmosphere; we will stabilize carbon levels when the flow of carbon into the atmosphere is balanced by the flow out. This is what is meant by net-zero carbon emissions. The stock of carbon in the atmosphere today is rising because the flow of carbon in exceeds the flow of carbon out. To stabilize the carbon in the atmosphere and reach our net-zero goals, we can reduce the flow in or increase the flow out or do a combination of both. Mitigation strategies, like reducing fossil fuel use, reduces the flow in. CDR strategies, as discussed in subsequent chapters of this book, increases the flow out.

Over 99.9 percent of the carbon on earth is in the lithosphere, which is the solid outer part of the earth made up of rocks and minerals. Most of that carbon is in an oxidized state (i.e., carbonate rocks), with only a tiny fraction (<0.01 percent) in the form of fossil fuels. After

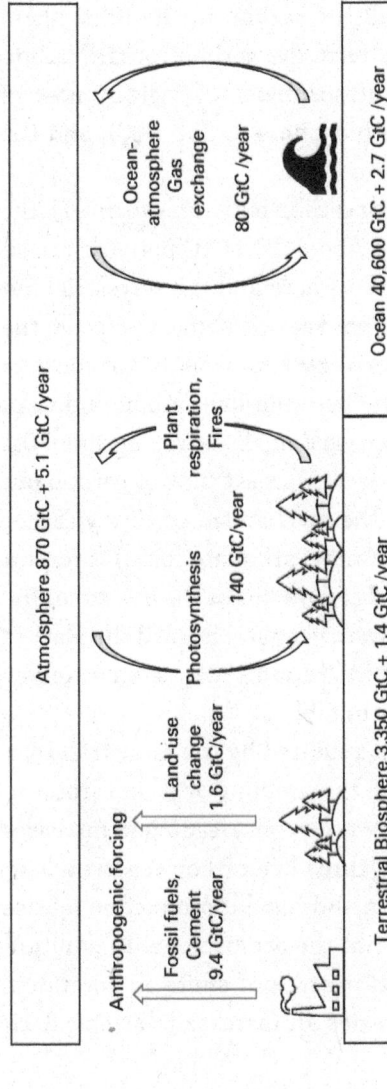

Figure 2 A simplified schematic of the global carbon cycle. The stocks and flows are indicative of the year 2019, and all values have some uncertainty. Note the large annual flows of carbon exchanged between the atmosphere and the ocean, between the atmosphere and the terrestrial biosphere, as well as the anthropogenic forcing from fossil fuel combustion, cement manufacture, and land-use change. Also shown are the stocks in the three live carbon reservoirs and their annual change.

Text within the figure:

Atmosphere 870 GtC + 5.1 GtC /year

Anthropogenic forcing

Fossil fuels, Cement 9.4 GtC/year

Land-use change 1.6 GtC/year

Photosynthesis 140 GtC/year

Plant respiration, Fires 140 GtC/year

Ocean-atmosphere Gas exchange 80 GtC /year

Terrestrial Biosphere 3,350 GtC + 1.4 GtC /year

Ocean 40,600 GtC + 2.7 GtC /year

H. E. Dunsmore,[2] we call the carbon in the lithosphere *dead* carbon. We use the term *live* carbon for the carbon distributed among the atmosphere (870 gigatonnes of carbon, GtC), terrestrial biosphere (3,350 GtC), and the ocean (40,600 GtC).[3]

There are large flows of carbon between the three large reservoirs of live carbon. About 140 GtC per year are exchanged between the atmosphere and the terrestrial biosphere, with about 80 Gt per year exchanged between the atmosphere and the ocean. By comparison, for most of recorded history, the flow between live carbon and dead carbon reservoirs has been very small, averaging about 0.2 GtC per year. Volcanism is the largest flow of carbon out of the dead zone, while the mechanism of rock weathering (e.g., the formation of new carbonate rocks) is responsible for the major flow of carbon out of the live zone. The stocks in each reservoir were very stable until the start of the industrial revolution in around 1750, as indicated by the small annual changes in table 2.

The use of fossil fuels required by the industrial revolution started to increase the amount of dead carbon released into the live carbon reservoirs, leading to increases in the carbon stock of all three live carbon reservoirs. Today, fossil fuel combustion and cement production release about 9.4 GtC per year into the atmosphere. In addition, land-use change (e.g., deforestation) emits an additional 1.6 GtC per year. This causes an increase in atmospheric

Table 2 Annual changes to carbon stocks (gigatonnes of carbon per year)

	Pre-industrial (natural carbon cycle)	Today (natural and anthropogenic)
Lithosphere	0.2	−9.2
Atmosphere	0	5.1
Terrestrial biosphere	−0.4	1.4
Oceans	0.2	2.7

carbon of about 5.1 GtC per year, with the rest going into the ocean (2.5 GtC per year) or the terrestrial biosphere (3.4 Gt per year). Without the land and ocean sinks, atmospheric CO_2 concentrations would be much higher today, exacerbating the impacts of climate change. Table 2 shows the annual changes of carbon stocks today in each of these reservoirs due to the combination of the natural carbon cycle plus the anthropogenic carbon emissions.

The Ocean Sink

A simple model of the ocean (see figure 3) helps explain its role in the global carbon cycle. This model divides the ocean into three layers. On top is a well-mixed *surface layer*, which is where CO_2 exchange with the atmosphere takes place. The next layer is called the *thermocline*. This layer

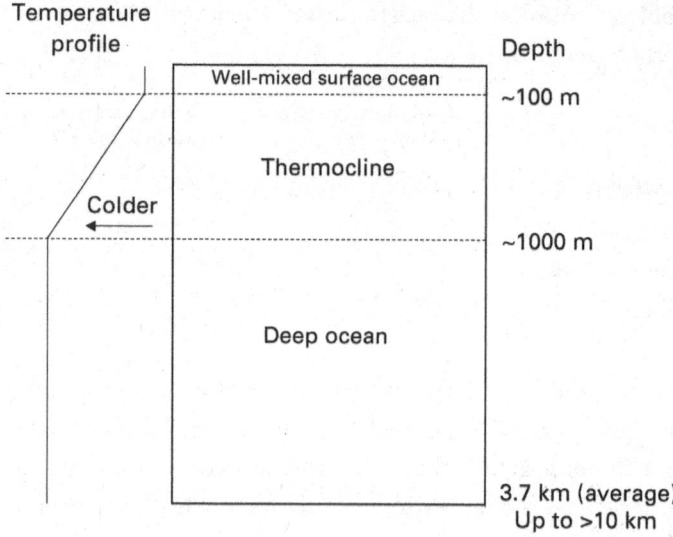

Figure 3 A simplified model of the ocean. Note that depending on location, the depths of the individual layers—surface layer, thermocline, and deep ocean—will vary.

separates the warm surface water from the cold waters of the deep ocean. Since cold water is denser than warm water, as the temperature decreases with depth the water density increases, making the thermocline stably stratified. This stratification acts as a barrier between the surface waters and the deep ocean, which inhibits vertical mixing. The thickness of these layers varies over time and space in the ocean. Typical thickness of the surface layer is 50 to 100 meters but can go up to several hundred meters.

The thermocline can be up to 1,000 meters thick. The deep ocean stretches from the thermocline to the *ocean floor*, which is about 3.7 kilometers deep on average but is over 10 kilometers deep in some places.

Most of the carbon in the ocean is inorganic. The surface layer of the ocean contains about 900 GtC, compared to 37,300 GtC in the thermocline and deep ocean. The CO_2 reacts with the water to form carbonate and bicarbonate ions. This chemistry greatly increases the solubility of CO_2 compared to other gases in water. For example, CO_2 is about thirty times more soluble than oxygen in seawater.[4] The exact partitioning of carbon between CO_2, carbonate ions, and bicarbonate ions depends on factors such as water temperature and salinity, but generally the bicarbonate ion accounts for about 85 percent of the inorganic carbon. In addition to this inorganic carbon, there is about 700 Gt of dissolved organic carbon and 1,750 GtC in the ocean floor sediments.

Approximately 80 GtC per year are exchanged in each direction between the atmosphere and surface ocean. CO_2 is outgassed from the oceans mainly in the tropics and high-latitude Southern Ocean, while CO_2 uptake occurs primarily in the mid-latitudes. Currently, there is a net flow of 2.5 GtC per year from the atmosphere into the ocean. The storage of this carbon in the ocean is a two-step process. First the CO_2 flows from the air to the surface ocean, and next the carbon flows from the surface ocean to the deep ocean.

As atmospheric concentrations of CO_2 rise, so does the amount of carbon in the surface ocean. Over a time-scale of months, the amount of CO_2 in the surface ocean will equilibrate with the concentration of CO_2 in the atmosphere. The equilibrium does have a temperature dependency because CO_2 is more soluble in cold water than warm water.

Transport of carbon from the surface ocean to the deep ocean is accomplished by two different mechanisms. The first is the *solubility pump*, where the cold ocean waters at high latitudes in the Arctic and Antarctic are dense enough to sink into the deep ocean. This phenomenon is also what drives the ocean currents in what is called the *thermohaline circulation*. The concentration of carbon in the cold waters is about twice the concentration of carbon in the warmer equatorial waters.

The second mechanism is the *biological carbon pump*. Via photosynthesis, the phytoplankton in the surface ocean will convert the CO_2 to organic matter. While most of this organic material will break down in the surface layer, about 20 percent will sink into the deep ocean. There most of it will be converted back to CO_2, with a very small fraction of the organic material being deposited on the ocean floor.

Transport of carbon from the surface waters to the deep ocean is slow, taking up to a millennium to equilibrate. Therefore, if we stop emitting all CO_2 to the atmosphere

If we stop emitting all CO_2 to the atmosphere today, the oceans would continue to be a CO_2 sink for centuries to come. Eventually, over 80 percent of the CO_2 we emit to the atmosphere today will end up in the ocean.

today, the oceans would continue to be a CO_2 sink for centuries to come. Eventually, over 80 percent of the CO_2 we emit to the atmosphere today will end up in the ocean. However, because of the slow transport of carbon to the deep ocean, the ocean is absorbing only 2.5 GtC per year of the 11 GtC per year being emitted to the atmosphere right now.

The Terrestrial Biosphere Sink

The carbon stocks in the terrestrial biosphere can be grouped into three main categories—vegetation (450 GtC), soils (1,700 GtC), and permafrost (1,200 GtC). The carbon flows between the terrestrial biosphere and the atmosphere are about 140 GtC per year in each direction. A major driver of these flows is vegetation, which removes CO_2 from the atmosphere through photosynthesis and emits CO_2 through respiration.

The impact of vegetation on CO_2 flows can be seen in figure 4. This is a graph of atmospheric concentrations of CO_2 that have been monitored very accurately since March 1958 at the Mauna Loa Observatory in Hawaii. The graph shows the steady increase of CO_2 levels over the decades (solid black line) and the monthly averages, which highlight the seasonal cycles (gray jagged line). During the growing season, photosynthesis lowers the atmospheric

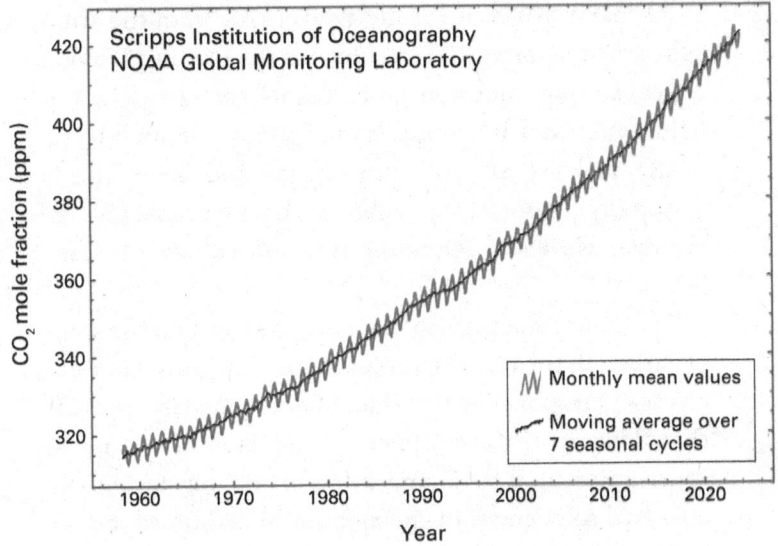

Figure 4 Atmospheric CO_2 concentrations measured at the Mauna Loa Observatory in Hawaii from 1958 to January 5, 2024. *Source:* "Atmospheric CO_2 at Mauna Loa Observatory," NOAA, accessed January 31, 2024, https://gml.noaa.gov/ccgg/trends.

CO_2 concentration, indicated by the downward slopes in the gray line. When the growing season ends, the gray line slopes upward until the growing season resumes again in the spring. Notwithstanding the seasonal ups and downs, the CO_2 concentration in the atmosphere is increasing at about 2 parts per million per year, corresponding to the 5.1 GtC per year increase in the carbon stock in the atmosphere.

A major driver of the net flow of CO_2 from the atmosphere to the terrestrial biosphere is the CO_2 fertilization effect. During photosynthesis, plants convert CO_2, sunlight, and water into sugar and starches. Elevated levels of CO_2 make some plants grow bigger and faster. This is why many commercial greenhouses have elevated CO_2 levels. Rising levels of CO_2 in the atmosphere have the same impact.

Land-use change also has a major impact on the stock of carbon in the terrestrial biosphere. Currently, land-use change is responsible for a loss of about 1.6 GtC per year that flows to the atmosphere. Deforestation is a major contributor to this. Changes in surface vegetation can also lead to changes in the amount of carbon contained in the underlying soils. Since the soils contain several-fold more carbon than the vegetation, understanding the impacts of these land-use changes on the soils is critical. For example, wetlands contain substantial amounts of carbon in their soils. So turning a wetland into cropland or a forest may significantly reduce the total carbon stock of that land.

There is much concern about the fate of carbon stocks tied up in the permafrost. Historically, the permafrost region has been a carbon sink. However, with rising temperatures and the melting of some permafrost, some scientists believe that the permafrost region is now a net source of carbon flowing to the atmosphere. Moving forward, there

Historically, the permafrost region has been a carbon sink. However, with rising temperatures and the melting of some permafrost, some scientists believe that the permafrost region is now a net source of carbon flowing to the atmosphere.

is concern that these carbon emissions from the permafrost will increase as the world warms. However, there is significant uncertainty in the magnitude and timing of these emissions.

Carbon Dioxide Removal and the Carbon Cycle

Emission reductions and carbon removals can be viewed as two sides of the same coin. From the viewpoint of the atmosphere, it makes no difference if you reduce your carbon emissions by one tonne or if you remove one tonne of carbon from the atmosphere as long as they happen at the same time. In either case, the atmosphere sees a reduction of one tonne of carbon relative to business as usual. However, there is a caveat in that you cannot magically either reduce your emissions or remove carbon. To reduce emissions or remove carbon, action needs to be taken, and it will have a carbon footprint attached to it. Therefore, one needs to look at the life-cycle emissions of the actions taken to cut emissions or remove carbon. Life-cycle emissions can vary greatly over the various emission reduction and carbon removal technologies. For carbon removal, a key life-cycle emission is associated with the energy used to remove the carbon. *Gross* carbon removed is the actual amount of carbon removed from the atmosphere, while *net* carbon removed is the gross carbon removed minus the life-cycle carbon emissions of the carbon removal technology. It is the *net* carbon removed that is the correct metric by which to assess a carbon removal technology.

We currently have net-positive carbon emissions, meaning that we are putting more carbon into the atmosphere than we remove. As long as we remain net positive, one unit of net carbon removed will have the same impact as one unit of net carbon emissions avoided. Until we get to net zero, this is a simple and appropriate way to look at the role of CDR.

At net zero, the land and ocean sinks will continue to take up CO_2 from the atmosphere but at a continually diminishing rate. The land sink will function for decades after net zero is reached, while the ocean sink will continue for centuries. The impacts of climate change will stop increasing on various timescales. Temperature rise will halt in a matter of years, permafrost melting will cease over decades, while sea-level rise will take centuries to stop.

Once we reach net zero, the driver for any additional carbon removal will be to lower the atmospheric stock of carbon—that is, to go net negative. Because of the carbon cycle, we cannot assume that if we remove a tonne of carbon from the atmosphere, the carbon stock in the atmosphere will decrease by a tonne. Removing that tonne will impact the carbon cycle by reducing the carbon flows from the atmosphere into the ocean and terrestrial biosphere. As a result, effectively reducing atmospheric stocks by a

At net zero, the land and ocean sinks will continue to take up CO_2 from the atmosphere but at a continually diminishing rate.

tonne of carbon will require removing multiple tonnes of carbon from the atmosphere.

Chapters 4 to 7 look at various CDR pathways and the ways they work with the carbon cycle to remove CO_2 from the atmosphere. One strategy is increasing carbon stocks in the terrestrial biosphere (chapter 4) and the ocean (chapter 7). Another strategy uses biomass as a "carbon pump" for removing CO_2 from the atmosphere and then harvesting and processing the biomass with storage of the carbon in the biosphere or lithosphere (chapter 5). One more strategy is to develop engineered systems that move carbon from the live reservoirs to the dead reservoirs (chapter 6). These pathways involve returning carbon to underground geologic formations (which include the same type of formations from which fossil fuels are extracted) as well as incorporating the carbon into rocks.

THE ROLE FOR CARBON REMOVAL

Carbon dioxide removal (CDR) is one of many tools available to address climate change concerns. None of the tools available, including CDR, is a silver bullet. We need to implement technologies and other approaches on many fronts. However, CDR is distinctly different from emission reduction technologies. Both can slow the rise of carbon dioxide (CO_2) stocks in the atmosphere and eventually stabilize atmospheric CO_2 concentrations, thereby stopping the impacts of climate change from getting worse. However, only CDR can roll back the clock and reduce the climate change impact of historical emissions, referred to as *legacy emissions*.

In this chapter, we first look at the conventional wisdom regarding the role of CDR in a portfolio to address climate change concerns. CDR can be used to compensate for current greenhouse gas (GHG) emissions and also to remove legacy emissions. We then look at why CDR is sometimes called a "moral hazard" or "dangerous distraction." Finally, we look at what happens once we reach a point of net-zero emissions and look at the possibility of using CDR technologies to go net negative.

Conventional View of the Role for Carbon Dioxide Removal

Figure 5[1] is one of innumerable possible scenarios of trajectories for greenhouse gas emissions and CDR deployment for the rest of this century. It is not a prediction but is used to illustrate the conventional view of the role for CDR. Emissions are the dark shaded area, while the amount of carbon removed from the atmosphere by CDR is the light shaded area. Emissions are strongly reduced over the next few decades but start leveling off, leaving us with what is commonly termed *residual emissions*. CDR is initially deployed in the late 2020s and continues to grow in magnitude for several decades. The solid line is the net CO_2 emissions to the atmosphere, which is shown to reach net zero in 2060. Before 2060, the emissions are net positive, and after 2060 they are net negative.

Residual emissions are emissions that are particularly hard and/or costly to eliminate. There will probably be some residual emissions in all sectors of the economy,

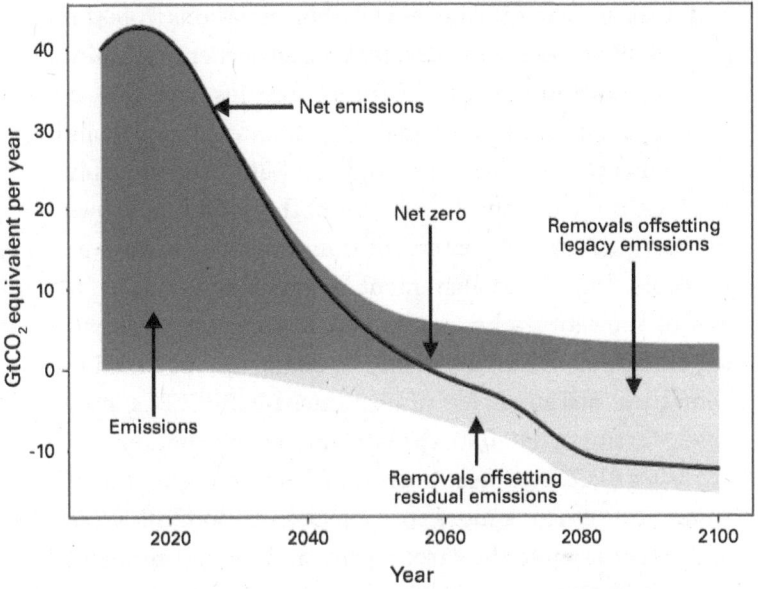

Figure 5 A scenario of greenhouse gas emissions and carbon removals from the year 2020 through 2100. Emissions are in dark gray, and removals are in light gray. The black solid line shows net emissions. In this scenario, net zero is reached in about 2060.

but the sectors most associated with residual emissions are agriculture, industry, and transport. Examples of agricultural GHG emissions that are hard to abate are nitrous oxide (N_2O) emissions from fertilizer use and methane emissions from livestock and rice paddies. In transport, emissions from airplanes, ships, and long-haul freight are expensive to abate. Industrial processes in cement, iron and steel, and chemicals all have carbon emissions from the process as well as energy use, making abatement difficult.

As shown in figure 6, CDR technologies have the potential to cap the cost of meeting climate change goals. While this example is from a study on BECCS, it can apply to all CDR technologies. In figure 6, the solid line shows the marginal cost of abatement if no carbon removal was available. The cost of abatement keeps rising in the latter half of the century because we must eliminate the residual emissions, which are costly to abate. The dotted line shows the marginal cost of abatement if BECCS is available. Starting after 2050, the two lines diverge because we are using BECCS to offset the residual emissions instead of eliminating them, a more cost-effective option. BECCS is effectively capping the carbon price at $240 per tonne of CO_2, which by the year 2100 is an order of magnitude less than the marginal cost without BECCS being available. The key to this type of behavior is to have an adequate supply of the technology at a relatively fixed price. The hope is that CDR technologies as a group can supply enough carbon removals at an affordable price to significantly reduce the cost of achieving net zero.

How much residual emissions will there be? In the literature, projections for CDR deployment in the year 2100 cover a wide range from 2 to 20 gigatonnes of CO_2-equivalent per year. The scenario in figure 5 shows residual emissions toward the lower end of that range. However,

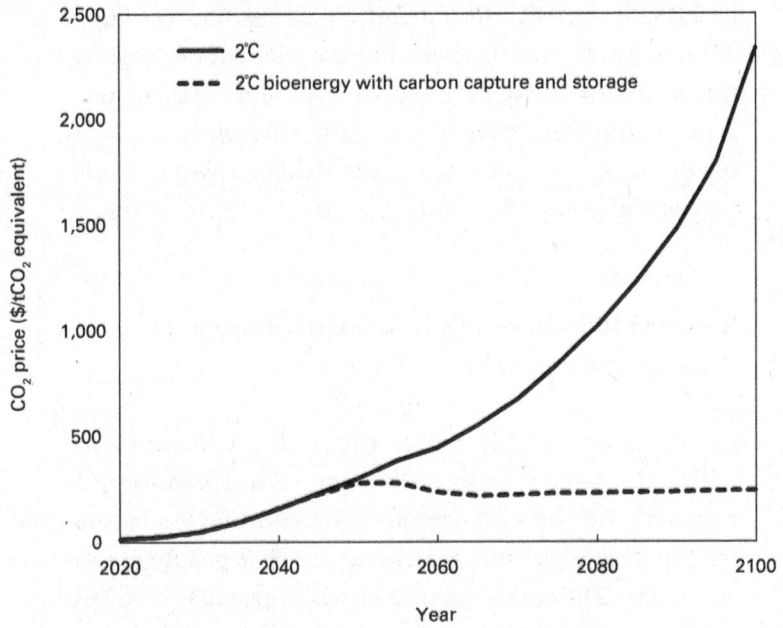

Figure 6 Scenarios of abatement costs to meet a 2°C stabilization target. Bioenergy with carbon capture and storage (BECCS) effectively caps the carbon price at about $240 per tonne of CO_2 equivalent, an order of magnitude less than the price in 2100 without BECCS. *Source:* Mathilde Fajardy, Jennifer Morris, Angelo Gurgel, Howard Herzog, Niall Mac Dowell, and Sergey Paltsev, "The Economics of Bioenergy with Carbon Capture and Storage (BECCS) Deployment in a 1.5°C or 2°C World," *Global Environmental Change* 68 (2021): 5, https://doi.org/10.1016/j.gloenvcha.2021.102262.

by 2100, it shows CDR contributing 20 gigatonnes of CO_2 net removal per year because we have gone net negative. Is this scenario realistic? It is hard to say today because there is much uncertainty about how CDR will scale in the coming decades. Therefore, speculation about how big a role CDR will play later this century is just that, speculation.

Is Carbon Dioxide Removal a Moral Hazard or a Dangerous Distraction?

In articles about CDR in both the academic literature as well as the popular press, CDR is sometimes branded as a "moral hazard" or a "dangerous distraction."[2] One reason for this characterization is the worry that people are focusing on CDR at the expense of reducing emissions. Put another way, CDR is being used as a *substitute* for emission reductions as opposed to a *supplement* to emission reductions. This is a legitimate concern that is fueled by real events, such as corporate announcements of plans to offset a significant part of their emissions (say, 50 percent) as opposed to deep cuts (say, 80 percent or more) in their emission levels. The situation is made worse when companies buy cheap offsets that are subsequently found to be of dubious quality, leading to charges of "greenwashing."

As stated previously, from a climate perspective, avoiding a tonne of CO_2 emissions to the atmosphere has the same impact as removing a tonne of CO_2 from the atmosphere. The key metric is the total stock of CO_2 in the atmosphere. Therefore, since they have similar impacts, it is important to understand whether is it cheaper to reduce CO_2 emissions by a tonne or to remove a tonne of CO_2. There is a wide range of costs for both CDR and emission reductions. However, today the most cost-effective portfolios for controlling atmospheric CO_2 concentrations would be dominated by emission reductions, not CDR. As we approach net zero, that will change. The cost of emission reductions will be much larger than today because we will implement the lower-cost emission reduction pathways first, leaving only the more expensive pathways. As a result, CDR will be called upon to play a much bigger role in getting us the final miles to net zero.

While the biggest role for CDR is in the future, there are important reasons to be working on these technologies today, the most important of which is that it may take decades to develop a technology to commercial maturity. Therefore, we must be working on CDR pathways today to have them commercially available when we need them in the future. Also, some CDR pathways are relatively inexpensive and economically attractive today. Since the goal is to reduce the rise of CO_2 stocks in the atmosphere as quickly as possible, these CDR options should be welcomed as a supplement to our emission reduction pathways.

Another major objection to CDR is its relationship to fossil fuels. Since the overwhelming amount of CO_2 emitted to the atmosphere comes from fossil fuel combustion, there is a worry from some quarters that implementing CDR implies continued use of fossil fuels. For example, the executive director of the nonprofit environmental organization Carbon180 has said, "The role of carbon removal in addressing climate change is to remove legacy emissions. It's not to offset continued fossil fuel use."[3] Once again, the atmosphere does not know or care whether the CO_2 is removed for legacy purposes or as an offset of fossil emissions. The relationship between CDR and fossil fuels depends on how climate policies evolve and how different players in the economy act. The two examples below show two very different views of the relationship between fossil fuels and CDR.

One of the direct air capture (DAC) companies, Climeworks (see chapter 6), has called for a clear and distinct role for CDR. Its proposal includes the following points:

- Besides emission reductions, CDR has an important role to play in the fight against global warming.

- More importantly, it has a *different* role to play and should not be substituting emission reductions.

- Hence, emission reductions and CDR should be clearly distinguished from each other—in climate pathways, target setting as well as in industry standards.

• A clear distinction is moreover needed in marketplaces and certificates generated from the two activities: whilst they are important and meaningful for the achievement of a net-zero world, credits generated from emissions reductions and avoided emissions should cease to exist, as soon as a net-zero state is achieved. At the same time the world will continue to rely on CDR markets to neutralize residual and historic emissions to maintain net-zero CO_2, and later on realize net-negative CO_2 emissions globally.

• Further, by explicitly splitting the contributions from emission reductions and removals, moral hazard is addressed, namely the claim that investing into CDR today could distract from emission reductions.[4]

At this time, Climeworks has not accepted any investments from the fossil fuel industry and has stated it does not intend to accept any in the future.

The other major DAC company, Carbon Engineering, has taken a very different path. It partnered with Occidental Petroleum Corporation (Oxy) to form a joint venture called 1PointFive to build a 500,000 tonne per year CDR plant. The company's CEO, Vicki Hollub, has stated, "We believe that our direct capture technology is going to be the technology that helps to preserve our industry over time. This gives our industry a license to continue to operate for the 60, 70, 80 years that I think it's going to be very much needed."[5] In August 2023, Oxy acquired all of Carbon Engineering for a purchase price of $1.1 billion.

Our (the authors') view is that the real issue is climate change and the ways that we can get to net zero as quickly as possible (going net negative is discussed in the next section). The goal is to reduce net CO_2 emissions, not to eliminate the use of fossil fuels. Because fossil fuel combustion is the major emitter of CO_2, this means that our use of unabated fossil fuels must be reduced significantly. However, even after reaching net zero, there can still be a role for fossil fuels in our energy systems. In all probability, the most cost-effective pathway to net zero will include some use of unabated fossil fuels.

As with many things, technology can be used for positive outcomes or negative outcomes, and CDR technologies are no different in this respect. To have positive outcomes, there must be good governance to ensure that removals are real and "permanent" and accounted for correctly. Ideally, as long as it helps with reaching climate stabilization goals, economics should be a major driver in the deployment of CDR. In other words, where it is less expensive to compensate for emissions by removing CO_2 than by eliminating those emissions, then it would be an

As with many things, technology can be used for positive outcomes or negative outcomes, and CDR technologies are no different in this respect.

appropriate use of CDR, not a moral hazard or a dangerous distraction.

Going Net Negative

The need to go net negative to correct an overshoot of temperature is an argument frequently made to justify the need for CDR technologies. As discussed in chapter 1, temperature rise is a function of atmospheric CO_2 concentrations. To meet a specified temperature rise (say, 1.5°C or 2°C), there is a limit to how much CO_2 emissions are allowed, termed the *carbon budget*. If we do not reach net-zero emissions in time but instead exceed this carbon budget, then there will be an overshoot of the desired temperature rise. One way to correct for this overshoot is by going net negative, meaning removing more CO_2 from the atmosphere than is being emitted.

While it may be desirable to go net negative at some point in the future, we feel it is a distraction to argue the need at this point in time. First of all, we cannot go net negative until we reach net zero, which will not happen for decades. There already is plenty of justification to pursue CDR technologies to help us reach net zero. When we do approach net zero many decades in the future, we will have much more information about the cost and performance of the various CDR technologies. We will also better understand the impacts of climate change and the benefits to be gained in going net negative. That is the time to have serious debates about whether going net negative is beneficial and, if so, how we can best to go about achieving it. Today, the focus should be on reducing emissions as quickly as practical and developing CDR technologies to operate at scale.

Some companies have publicly stated net-negative goals. By going net negative, a company can, over time, offset all of its legacy emissions and claim that it is now carbon neutral since its inception. The use of the term *net negative* in this instance is very different from its use in global emissions. Today, global emissions are net positive, so whether a reduction in net CO_2 emissions comes from emission reduction or from carbon removal is irrelevant to the carbon stock in the atmosphere. However, one can question the cost-effectiveness of paying hundreds of dollars per tonne of CO_2 for CO_2 removal versus paying tens of dollars per tonne of CO_2 for emission reduction. For example, for the same amount of money, setting up a program to insulate housing for low-income families would yield much greater net CO_2 emission reductions than buying expensive carbon removal credits. It would also have the immediate benefits of reducing heating and cooling bills for those families. However, from a marketing point of view, opting for reductions may not be as catchy as saying the company is now carbon neutral since its inception.

ENHANCING LAND SINKS

Enhancing the carbon stocks in the land and ocean can result in a net removal of carbon from the atmosphere. In this chapter, we look at strategies and methods for enhancing the carbon stocks in the land (chapter 7 looks at enhancing the ocean sink). The technologies and strategies proposed for enhancing the land sink are also referred to as *natural climate solutions*, as *agriculture, forestry, and other land uses (AFOLU)*, and as *nature-based solutions*. These strategies can be categorized as follows:

• *Protect.* Currently, land-use change emits 1.6 gigatonnes of carbon (GtC) per year into the atmosphere. Protecting land aims to reduce this carbon flow.

• *Restore.* Restoring degraded ecosystems back to their natural state can increase their stock of carbon, thus resulting in the net removal of carbon from the atmosphere.

• *Manage.* This covers a wide range of strategies by which humans transform and/or manage an ecosystem to increase its carbon stocks.

While all three strategies enhance the land sink, only restore and manage can be considered carbon dioxide removal (CDR) pathways. Protect is not a CDR pathway as it avoids carbon emissions to the atmosphere as opposed to removing carbon from the atmosphere.

While the above strategies sound straightforward in theory, they pose many challenges in implementation. These challenges are briefly defined below and are discussed in more detail in chapter 8:

• *Additionality.* Are the net carbon removals from the execution of a project more than (i.e., additional to) the net carbon removals that would have happened if the project had not taken place?

• *Permanence.* Will the carbon removed remain out of the atmosphere for a sufficiently long period?

• *Leakage.* Will the execution of a project to increase the carbon stocks of a certain parcel of land lead to the release of carbon stocks from a parcel of land

The technologies and strategies proposed for enhancing the land sink are also referred to as *natural climate solutions*, as *agriculture, forestry, and other land uses (AFOLU)*, and as *nature-based solutions*.

somewhere else? For example, if land is protected from unsustainable logging, will loggers simply find another piece of land to log?

• *Monitoring, reporting, and verification (MRV)*. Can the increase in carbon stocks be accurately measured and verified?

• *Environmental impacts*. What are the environmental impacts (such as the effect on biodiversity) of the project?

• *Environmental justice*. Does the project represent ethically and socially responsible deployment?

Below, the strategies and projects for enhancing the land sink are organized into three groupings—forests, croplands and grasslands, and wetlands.

Forests

In the Australian outback, carbon farming has taken off with the planting of mulga trees. In northern New South Wales alone, about 150 properties have earned over $300 million in the last decade from carbon credits.[1] A San Francisco start-up, Living Carbon, has bioengineered a poplar tree to hold up to 27 percent more carbon.[2] The

World Economic Forum launched 1t.org, whose mission is "conserving, restoring, and growing one trillion trees by 2030."[3] What do all these efforts have in common? They all want to increase the amount of carbon stocks in the terrestrial biosphere. In other words, they want to enhance the natural land sink.

Forestry projects can be classified as *reforestation*, *afforestation*, or *improved forest management*. According to the Intergovernmental Panel on Climate Change (IPCC), "Reforestation and afforestation can be defined as the conversion of non-forested lands to forests with the only difference being the length of time during which the land was without forest," with reforestation having a shorter time without forests. It further states that "Afforestation is usually defined as the establishment of forest on land that has been without forest for a period of time (e.g., 20–50 years or more) and was previously under a different land use."[4] Improved forest management strategies are applied to existing forests and can include "lengthening harvest schedules, thereby generally increasing the age and carbon storage of the forest on average; improved fire management; thinning and understory management; and improved tree plantation management."[5]

Most people would consider the planting of more trees as a very positive activity, not just for carbon storage but for a whole host of other reasons, including maintaining biodiversity, improving water resources and air quality, and addressing sustainability goals by empowering local communities. When done right, these projects produce many benefits at a relatively small cost. However, planting trees at a large scale has some important issues that must be addressed. Bonnie G. Waring and her colleagues conducted an expert elicitation survey regarding concerns about forest-based natural climate solution projects.[6] The survey highlighted concerns about "replacement of biodiversity-rich habitat with 'carbon plantations,' and increased vulnerability of forest carbon to future disturbance." In other words, replacing intact forests with fast-growing tree monocultures can harm biodiversity and also weaken the resiliency of the forest to wildfires, pestilence, floods, and droughts.

Another issue is termed *land-use change*. This comes in two flavors—direct land-use change and indirect land-use change. *Direct land-use change* refers to climate impacts that a forestry project may have beyond just the increase in the carbon stock of the trees. For example, will the project cause changes in the carbon stock of the soils? Another example is the change in the albedo of the land. The *albedo* refers to how reflective the land is. The dark canopy of a forest may absorb more solar radiation than, say, a grassland, leading to more warming. This impact varies geographically, "with Earth system models [showing] that an expansion of forest in the tropics would result in cooling, while afforestation in the boreal zone [the climate zone south of the Arctic] might have only a limited effect or might even result in global warming. There are also significant uncertainties about the impacts on non-carbon dioxide greenhouse gases, emissions of volatile organic compounds, evapotranspiration (the combination of evaporation from land and transpiration from plants), and other issues that can influence the net climate effects of forestry projects."[7]

Indirect land-use change refers to the climate impact of a project on land outside the project's physical footprint. There is competition for land for many purposes, including growing crops, grazing livestock, producing forest products like timber and pulp, and developing residential and commercial properties. If a project converts land that was used for grazing into a forest, does that induce a land-use change elsewhere to replace the lost grazing land? If so, it can easily change a project that looks like it removes carbon to one that actually emits carbon. Unfortunately, indirect land-use change is very difficult to identify and quantify, creating uncertainty around the exact benefits of many forestry projects.

Beyond the climate impact of land-use change is the economic impact. For example, adding land to forestry may impact the price of croplands and therefore food. There is no consensus on how significant a problem this may be, with the literature showing a wide range from very little impact to significant impact. Another issue involves water resources. Many places in the world have water stress, and the list is growing due to climate change. Forestry projects can have either a positive or a negative impact on water quality and quantity, though the literature generally reports that forestry projects can help alleviate water stress.

Croplands and Grasslands

The objectives and issues for carbon sequestration in croplands and grasslands are similar to those in forests, except here the focus is more on storing the carbon in the soils as opposed to the vegetation. "The stock of carbon in the soil over time is determined by the balance between carbon inputs from litter, residues, roots or manure, and losses of carbon, mostly through microbial respiration and decomposition, which is increased by soil disturbance."[8] Compared to CDR projects in forests, projects to increase the carbon stocks in the soils of croplands and grasslands do not have as long a history and are currently an order of magnitude or two smaller in the amount of carbon stored.

The amount of carbon that can be stored in soils depends on the soil type, the meteorological conditions (e.g., rainfall and temperature), and land management techniques. Since these conditions are highly variable, there is no one-size-fits-all strategy for soil carbon sequestration. Some of the more prominent strategies being discussed today are described below.

No-till agriculture refers to growing practices for crops or agriculture that minimize disturbances to the soil. Not disturbing the soil minimizes carbon losses from decomposition and microbial respiration and allows the carbon stock in the soils to grow. Where tillage turns over the soil, no-till agriculture minimizes soil disturbance by making only a slit or holes in which to plant seeds. Some initial capital expenditures are required for items like specialized seeding equipment, but there are annual savings in labor and fuel because annual plowing is no longer required. Because no-till farming leaves crop residues in the fields, this increases water retention, which improves water conservation but can encourage fungal diseases. Since plowing disrupts weed growth, the absence of plowing can mean more herbicides are needed. No-till farming can lead to higher crop yields, but a significant learning curve is involved in its implementation.

Cover crops, such as grasses and legumes, are grown after the primary crops are harvested and when fields are normally bare, thereby providing carbon inputs to soils. It has been estimated that growing cover crops on the fields of the five primary crops in the United States not already using cover crops can result in about 100 megatonnes of carbon dioxide ($MtCO_2$) equivalent per year of removals. In addition, cover crops can improve soil health, yields, and yield consistency.[9] *Improved manure management* can increase soil carbon as well as reduce emissions of methane.

Improved nutrient management refers to the management of nitrogen fertilizers, which reduce "N_2O emissions due to more efficient use of nitrogen fertilizers and avoided upstream emissions from fertilizer manufacture." Improved nutrient management practices include "(i) reduced whole-field application rate, (ii) switching from anhydrous ammonia to urea, (iii) improved timing of fertilizer application, and (iv) variable application rate within field."[10]

Biochar is a charcoal that is made from biomass and added as an amendment to soils. Recently, interest in biochar has increased greatly because of its potential as a carbon removal technology. Biochar is discussed further in chapter 5 with other biomass-based carbon removal pathways.

Measuring the amount of carbon in the soil is not simple. It usually involves taking representative core samples of the soils and analyzing them for total organic carbon. This method is currently both time consuming and expensive. In Australia, farmers can get paid for increasing soil carbon, but the amount must be verified by measurements. Unfortunately, the cost of verification can exceed the payments for carbon sequestration. New techniques to measure and/or estimate the carbon content of soils are under development and, if successful, will help alleviate the measurement challenge.

Another challenging issue for soil carbon is permanence. Strategies to increase the carbon stock in soils practiced over time will eventually saturate the soils. At that point, the practices must be maintained, because the organic matter, which contains the carbon, can easily decompose. Ensuring that there will be no land-use change for a century or longer will be difficult enough. However, ensuring that soil management practices adopted for CDR purposes will be maintained for a century or longer is even more difficult. This has led some to quip that "soil carbon is hard to gain and easy to lose."[11]

Wetlands

Wetlands are important ecosystems because they can hold significantly more carbon per unit area compared to most other types of ecosystems. Wetlands are distinguished not by the amount of carbon in the vegetation but by the amount of carbon held in the soils. This makes restoring degraded wetlands an important CDR strategy. On the other hand, projects that propose to drain wetlands to grow a forest are not a good idea and will actually end up decreasing carbon stocks.

Peatlands, a type of wetland, comprise about 3 percent of the planet's land mass but contain almost 20 percent of the land carbon stocks. Near the North Carolina coast is a large peatland called a *pocosin*, a Native American word meaning "swamp on a hill." European settlers first logged

Wetlands are important ecosystems because they can hold significantly more carbon per unit area compared to most other types of ecosystems.

this land and then drained the peatlands so they could be farmed. These activities released large amounts of carbon from the land. They also made the land more vulnerable to fires, like the Alan Road fire that burnt 95,000 acres in the state in 1985. Five years later, in 1990, the US government established the Pocosin Lakes National Wildlife Refuge, which today is home to one of the largest wetlands restoration projects in the country. Over the past two decades, the project managers developed an ingenious hydrologic system that mimicked the original natural system in order to bring water back to the peatlands. Over time, this will restore the peatlands back to their natural state, which includes restoring their lost carbon stocks.[12]

Restoration, like the other natural climate solutions discussed in this chapter, removes carbon from the atmosphere over time. It may take decades or even more than a century for the carbon stock to be fully restored. Over that time period, the amount of carbon captured per year may vary. As the restored land becomes saturated with carbon, uptake rates will slow down and eventually cease. If a project wants to claim carbon credits (see the carbon market discussion in chapter 9), the time-varying nature of the carbon sequestration introduces some challenges. First is measuring the uptake of carbon over time, which as discussed above can be a significant cost. Second is identifying when the credits will be issued. Project developers would naturally want to be paid up front in a lump

Peatlands, a type of wetland, comprise about 3 percent of the planet's land mass but contain almost 20 percent of the land carbon stocks.

sum. However, it can be argued that it is more appropriate for the developer to be issued credits over time to reflect the actual carbon uptake. The difference in these two approaches will significantly impact project financing.

The term *coastal blue carbon* is commonly used to describe CDR projects in coastal wetlands. These CDR projects involve "land use and management practices that increase the organic carbon stored in living plants or soils in vegetated, tidally-influenced coastal ecosystems such as marshes, mangroves, seagrasses, and other wetlands."[13] Coastal wetlands are very vulnerable to climate change, especially as sea levels rise. Over time, some existing wetlands will be lost to the sea, while new wetlands will emerge as the sea moves inland. Managing these changes will be a key priority for coastal wetland management.

BIOMASS-BASED CARBON REMOVAL AND STORAGE

In the previous chapter, we looked at carbon removal by enhancing the land sink. The objective of many of those carbon removal methods was to increase the stock of biomass—and therefore carbon—stored on a given piece of land. In this chapter, we look at carbon removal methods that harvest and process the biomass to store its carbon in a variety of sinks. Another way to think of this is that the biomass acts as a "carbon pump" for removing carbon dioxide (CO_2) from the atmosphere and then subsequently storing the carbon in the biosphere or lithosphere. Over time, this has the advantage of removing much more carbon per unit of land compared to simply enhancing the land sink. However, in terms of dollars per tonne of carbon (C) removed, this will generally be more expensive than the methods discussed in chapter 4 because of the costs associated with the harvesting and processing the biomass as well as the storage of the carbon. Also, to be an effective carbon removal strategy, the carbon stocks on a given piece of land must be maintained. This means that the carbon in the harvested biomass must be balanced by the production of an equal amount of new biomass on the same piece of land.

Today biomass is harvested for many commercial products, including food, pulp and paper, and lumber. These products have a wide range of lifetimes, as short as days for some food products and as long as centuries for some lumber products. While long-lasting products made from lumber can be considered carbon removal, we do not consider this category in this chapter as this is a well-established business. Instead, we focus on two groupings of biomass-based carbon removal pathways that are being explored primarily to address climate change. The first grouping of pathways, bioenergy with carbon capture and storage (BECCS), harvests the biomass for its energy content as well as its carbon content. The three main energy products under consideration are electricity, hydrogen, and biofuels. The carbon captured from BECCS projects is generally in the form of CO_2, which can be stored in geological formations. The second grouping of pathways simply processes the biomass to a form in which it can be readily sequestered. Here, we examine three options— biochar, bio-oils, and biomass burial. We end this chapter with a discussion of sustainable biomass supply that can be harvested and processed to derive a carbon removal service and that is essential for all biomass-based carbon removal.

Bioenergy with Carbon Capture and Storage

Before fossil fuels, biomass (e.g., wood) was the major energy source for humanity. Today, while modern biomass supplies only 6 percent of the world's total energy, we still use more biomass in the provision of energy services today than we did in 1800. In terms of energy density, biomass is a poor fuel compared to the fossil alternatives, with an energy density in the range of 15 to 20 megajoules per kilogram (MJ/kg) (dry weight), compared to coal at 18 to 35 MJ/kg, gasoline at 45 MJ/kg, and natural gas at 50 to 55 MJ/kg. However, what drives the interest in BECCS is the value of the carbon, which, as we approach net zero, will be significantly greater than the value of the energy in the biomass. For example, assuming an energy density of 18 MJ/kg for biomass and a value of the energy at $3 per gigajoule (GJ) (which is greater than the January 2024 price for coal or natural gas), the energy value of biomass is $54 per tonne. Woody biomass is approximately 50 percent carbon, and taking the average carbon price in 2023 for the EU Emissions Trading System of about $100 per tonne of CO_2, the value of the carbon is $183 per tonne of

Before fossil fuels, biomass (e.g., wood) was the major energy source for humanity.

biomass. In this example, the value of the carbon is over three times greater than the value of the energy. As we approach net zero, this ratio will become even greater.

The first step in any BECCS process (see figure 7) is to source the biomass, which is a catch-all term that covers a range of potential sources of biomass that have different levels of availability, energy density, and sustainability implications. Roughly, these categories include the following:

1. Woody biomass (from forestry residues, sawmills, construction and demolition waste, old furniture, etc.)

2. Dedicated bioenergy crops, which are usually a variety of woody plants or grasses (such as willow or poplar trees and different types of elephant grass) and agricultural residues (the material leftover after harvest, such as straw, corn stover, rice husks, and bagasse)

3. Municipal solid waste (including domestic food waste) and commercial residues (such as used cooking oils)

4. Animal manure and sludge from wastewater treatment

Not all forms of biomass are equally suitable for every BECCS process. While waste and by-product biomass may be more cost-effective initially, its supply is limited, so at larger scales it is envisioned that dedicated bioenergy

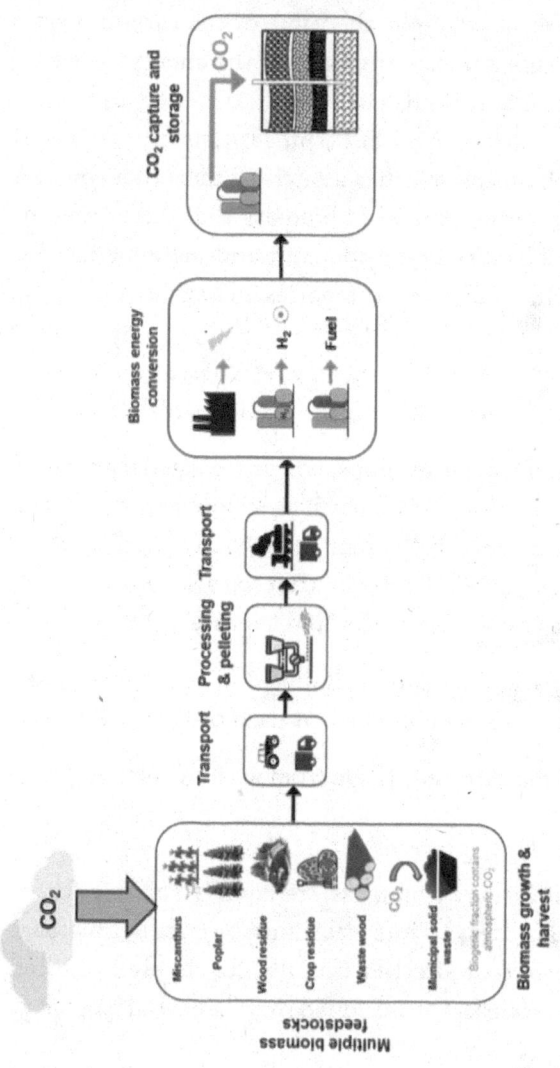

Figure 7   Illustration of the typical steps for a bioenergy with carbon capture and storage (BECCS) removal pathway.

crops will become the predominate source. More details are discussed below in the section on biomass sourcing.

The heart of a BECCS process is extracting energy from the biomass with the capture and storage of the associated CO_2 emissions. Since the biomass is made up of carbon that originated in the atmosphere, the CO_2 that is captured and stored can be counted as carbon removal as long as it is balanced by new production of biomass. A life-cycle assessment needs to be conducted to determine the net amount of carbon removed, as there may be greenhouse gas emissions associated with the production, harvesting, transport, and processing of the biomass. Any of the biomass CO_2 that goes back into the atmosphere is typically considered to be carbon neutral, meaning there is no penalty for it. You just cannot claim it as a carbon removal.

There are many ways to convert biomass to electricity or to a fuel. In the following subsections, we look at three of the most prominent pathways being investigated. Among other items, they vary in complexity, maturity, cost, and fraction of the biomass carbon captured. This section is followed by sections on the transport and storage of the captured CO_2.

Bioelectricity with Carbon Capture and Storage

Conceptually, bioelectricity with carbon capture and storage (CCS) is the simplest of the bioenergy with CCS (BECCS) processes. Biomass-fueled power plants exist today and are very similar to coal-fired power plants. The biomass is burned in a boiler that raises steam that is sent to a steam generator to produce electricity. There are far, far fewer biomass-fired power plants today than coal-fired power plants, for two main reasons. First, biomass is a poorer fuel, having both a lower energy density and a higher moisture content than coal. This results in significantly lower conversion efficiencies for electricity production from biomass. The conversion efficiency is the percentage of the energy content of the fuel that gets turned into electricity. Second, coal is generally a less expensive and more convenient fuel to use in power plants. As a result, biomass power plants exist mainly where there is a convenient, cost-effective supply of biomass. For example, biomass-fired power boilers are typically found at pulp and paper mills where a good supply of waste biomass is available.

In the boiler of the power plant, the biomass is combusted with air, creating a flue gas. By combusting the biomass, most of the carbon in the biomass gets converted to CO_2 in the flue gas, which is ultimately sent up a smokestack and vented to the atmosphere. In order to capture the CO_2 from the flue gas, the flue gas is sent to a carbon capture process prior to going up the smokestack. This has been demonstrated at two coal-fired power plants, each capturing over a million tonnes of CO_2 per year. The first was in 2014 at the Boundary Dam plant in Saskatchewan, Canada, followed in 2016 at the Petra-Nova plant near

Houston, Texas. The carbon capture process used in both plants is referred to as *amine scrubbing* and it can work just as well with biomass as with coal because their flue gases are similar. While no large-scale demonstrations of carbon capture from biomass power plants exist today, it has been demonstrated on a smaller pilot plant scale.

There are a number of ways to capture CO_2 from power plants,[1] but amine scrubbing is the only one being applied at a commercial scale today. In this process, the flue gas is sent to a large tower caller an *absorber*. The flue gas enters the bottom and rises up the tower. The amine, which is dissolved in water (typically, 20 percent to 30 percent by weight), enters the top of the tower and flows down. The tower is full of packing (picture crumpled aluminum foil), which is wetted by the amine and creates a large surface area for the flue gas to contact the amine solution. Upon contact, a chemical reaction occurs between the amine (a base) and the CO_2 (an acid), such that the CO_2 is chemically bonded with the amine in solution, thereby removing it from the flue gas. The CO_2-depleted flue gas exits the top of the tower and is then released to the atmosphere, while the CO_2-rich amine solution exits the bottom of the tower, is heated, and then sent to a second tower called a *stripper*. In the stripper, CO_2 is released from the amine solution and compressed to a liquid state ready for transport and permanent storage (described later in this section). The amine solution is then recycled back to the absorber.

Bioelectricity has significant advantages compared to other types of BECCS. It is based on commercially available technologies for both the power plant and the CO_2 capture plant. It is also possible to capture a very high percentage (over 95 percent) of the carbon in the biomass. This is important because it is the amount of carbon that is captured that really drives the economics. This has led this type of BECCS to be included in large energy-environmental models that inform climate change policy. The authors of this book were part of one such study, concluding that "We find that [bioelectricity with CCS] could make a substantial contribution to emissions reductions in the second half of the century under 1.5 and 2°C climate stabilization goals, with its deployment driven by revenues from carbon dioxide permits. Results show that global economic costs and the carbon prices needed to hit the stabilization targets are substantially lower with [BECCS] available, and BECCS acts as a true backstop technology at carbon prices around $240 per tonne of carbon dioxide."[2] Whether this type of BECCS can live up to model projections is still to be determined, but we remain optimistic for the prospects for bioelectricity with CCS as a major carbon removal pathway.

Biofuels with Carbon Capture and Storage

The world's transportation sector runs almost exclusively on liquid fuels derived from oil. Recently, there has been a push to electrify transportation—specifically, through the introduction of battery electric vehicles. However, not all parts of the transportation sector are easy to electrify, so it is anticipated that significant amounts of liquid fuels will be needed throughout this century. The question is how to produce carbon-neutral liquid fuels. The answer falls into three categories:

1. Continue to use oil-based liquid fuels, but offset their GHG emissions with negative emissions generated from carbon removal technologies.

2. Make liquid fuels, which are termed *biofuels*, from sustainably grown biomass.

3. Make liquid fuels, sometimes called *e-fuels*, by combining hydrogen made from electrolysis of water using carbon-free electricity with CO_2 from the atmosphere. The atmospheric CO_2 is obtained from a process like BECCS or direct air capture (see chapter 6).

While all three options require some type of carbon removal, only the making of liquid fuels from biomass has the potential to generate carbon removal credits. To achieve this potential, CCS must be part of the biofuels manufacturing process.

One challenge in generating net carbon removals from biofuels with CCS is that the liquid fuel created will contain carbon, which will return to the atmosphere when the fuel is burned. While the carbon originally came from the atmosphere and is therefore neutral from a climate perspective, it does mean that a much smaller percentage of carbon in the biomass is available for generating net carbon removals. This percentage can vary depending on the processes used, but generally it will be below 50 percent. Since the carbon in the biomass is a big economic driver, this greatly impacts the economics. On the other hand, liquid fuels are a high-value product, attracting a greater price than electricity, resulting in a positive economic impact.

There are quite a few pathways available for converting biomass to biofuels. The pathways can be categorized as either biological or thermochemical. The major *biological* pathway is fermentation, where bugs turn the biomass into ethanol and generate a stream of CO_2 as a by-product. This is being done today at large scale in the United States using corn as a feedstock and in Brazil using sugar cane. The ethanol is not used directly as a transport fuel but is blended with gasoline. Because the by-product CO_2 is very pure, it generally only needs to be compressed to be made ready for transport and storage. This is being done today at a number of ethanol plants in the United States, with plans to add CCS to more ethanol plants currently in the development stage (e.g., see discussion in the CO_2 transport section below). While this whole process reduces fossil CO_2 emissions from transportation fuels, the amount of negative emissions it generates, if any, is very small. This is because only about 20 percent of the carbon in the biomass ends up as by-product CO_2, setting a maximum for the carbon removal potential. After accounting for life-cycle CO_2 emissions from fossil fuel use associated with the process, in many cases there is no net carbon removal.

Thermochemical pathways, such as gasification or pyrolysis, are options for making biofuels. They have several advantages compared to the biological pathways. Thermochemical processes can use just about any biomass feedstock, whereas biological processes use primarily sugars and starches as feedstocks because much biomass (termed *cellulosic biomass*) contains components (like lignin) that are hard to break down biologically. Thermochemical processes can also produce *drop-in fuels* (such as biodiesel), which can substitute directly for existing fuels used by trucks, ships, and airplanes. The ability to produce drop-in fuels is important for industries like air transport, which has strict specifications for fuel, and altering the fuel specifications could require expensive modifications of aircraft. Finally, thermochemical processes have the potential for much greater economies of scale compared to biological processes. However, thermochemical processes are not as mature as biological processes and are more expensive today. Since about 2010, the US government has tried to encourage an increased use of cellulosic biomass to make biofuels through its Renewable Fuel Standard Program, but this effort has met with limited success because of the high costs of producing those biofuels.

Thermochemical processing to produce synthetic fuels has a long history dating back to the late nineteenth century, when coal was gasified to make "town gas" to be used for lighting and cooking. When Germany's access to conventional liquid fuels was cut off during World War II, the country made synthetic liquid fuels from coal via gasification, and more recently, South Africa did the same during its apartheid period. During the oil shocks of the 1970s, major synthetic fuels programs were funded by governments worried about access to oil and gas. All these efforts helped develop thermochemical processes to make liquid fuels, with the starting material usually being coal. What is different now is that we are interested in synthetic fuels in order to reduce the carbon footprint of oil and gas. Therefore, the interest is now to use biomass, not coal, as the feedstock to create synthetic fuels such as biofuels. However, as discussed below, biomass poses some additional challenges as a feedstock compared to coal.

While multiple thermochemical processes can create biofuels, here we discuss only gasification because it is the most common approach. The major steps involved are as follows:

1. Biomass is gasified at high temperatures and pressures and under oxygen-lean conditions to create

a *synthesis gas* (syngas) consisting primarily of carbon monoxide (CO), CO_2, water vapor (H_2O), and hydrogen (H_2). The exact process conditions depend on the type of gasifier. Fluidized bed gasifiers are most typical for biomass gasification today, but other options include entrained-flow and plasma gasifiers.

2. The CO to H_2 ratio can be adjusted through the water–gas shift reaction: $CO + H_2O = CO_2 + H_2$. For creating biodiesel, a H_2 to CO ratio of about 2 is desired.

3. CO_2 needs to be removed from the synthesis gas. Since the synthesis gas is at an elevated pressure, CO_2 can be removed more easily and cheaply compared to removing CO_2 from power plant flue gases using the amine scrubbing process.

4. A Fischer–Tropsch synthesis process can create larger hydrocarbon molecules from the clean synthesis gas. It uses a catalyst, requiring the synthesis gas to be free of impurities, such as unwanted hydrocarbon compounds that may form in some gasifiers. A synthetic oil results, which can be refined into biodiesel or aviation fuel.

The use of biomass as a feedstock to a gasifier poses some challenges. It has a high moisture content, requiring removal of most of the moisture prior to gasification. This can be accomplished via pelletization, which is a well-established process where the biomass is exposed to high temperatures and pressures to drive off the moisture and turn it into pellets. Certain gasifiers require the biomass to be finer particles, similar to pulverized coal. This can be accomplished through a process called *torrefaction*, a type of pyrolysis where the biomass is heated in the absence of oxygen to drive off its volatile matter. Pyrolysis is described further in the biomass-based CDR section later in this chapter. The resulting torrefied biomass resembles a charcoal that can easily be ground into small particles that can be fed to a gasifier. This preprocessing of the biomass contributes to making biomass gasification more expensive than coal gasification. The energy used to either pelletize or torrefy the biomass generally comes from the biomass itself. Studies[3] have shown that it is possible for about 40 percent of the carbon in the biomass to be counted as net carbon removal from the type of gasification processes described above.

In summary, there is a good chance we will need to be using liquid fuels in a net-zero world. If this is the case, carbon removal will likely play an important role. If we continue to use fossil-generated fuels, carbon removal can provide carbon offsets. If we generate e-fuels, a slipstream from the carbon removal process can provide the required CO_2 feedstock. Finally, if we produce biofuels, carbon removals generated from the process can provide revenue to improve process economics. However, if the goal of implementing BECCS is to maximize carbon removals from a fixed amount of biomass, then bioelectricity with CCS will be a more favorable option.

Biohydrogen with Carbon Capture and Storage

One of the problems with biofuels as described above is that they contain carbon that is released to the atmosphere when they are burned. This decreases the net amount of carbon removal that can be obtained from a given amount of biomass. However, there are carbon-free fuel options—specifically, hydrogen, either in its pure form (H_2) or as ammonia (NH_3). The market for hydrogen is expected to grow in a carbon-free world, but there are many approaches to making hydrogen, usually designated by colors. For example, hydrogen from renewable energy is labeled "green"; from nuclear energy, "pink"; and from natural gas with CCS, "blue." No color has been assigned to hydrogen made from sustainable biomass, probably because it is too expensive to compete today. However, if carbon removal can be monetized, then the economics could improve greatly for biohydrogen.

Hydrogen from biomass is generally produced via gasification, as described above for biofuels. Once the synthesis gas is made, the water–gas shift reaction is used to convert the CO to H_2 and CO_2 by reacting the CO with H_2O over a catalyst. Once the CO_2 is removed, the synthesis gas is sent to a commercially available *pressure-swing adsorption* (PSA) process to purify the hydrogen. Overall, this process is simpler than making liquid fuels from biomass. However, the market for liquid fuels today is many times greater than the hydrogen market, though that may change as we head to net zero.[4]

Carbon Dioxide Transport

The above bioenergy pathways capture CO_2, but the CO_2 must be permanently stored to complete the BECCS process. This is generally done in underground geologic formations (see the next section). At the site of the bioenergy process, the CO_2 will be compressed, which then makes it ready to be transported to the storage site.

Commercial technologies are readily available from vendors for CO_2 transport. Tanker trucks regularly transport tens to hundreds of tonnes of CO_2 for distances up to 300 km. For the millions of tonnes of CO_2 per year quantities that are expected to be associated with BECCS, pipelines are the preferred mode, with ship transport a possible alternative under certain circumstances.[5]

In the United States, over 6,500 km of CO_2 pipelines have been built primarily to support *enhanced oil recovery* (EOR) operations. To ensure single-phase flow and to maintain high densities in the pipeline, the CO_2 pressures are kept above the CO_2 critical pressure of 73.9 bar.

To protect against corrosion, water is removed to keep its concentration less than 50 parts per million (ppm). Pipeline transport scales well; doubling the pipe diameter increases the pipeline capacity by a factor of four. A 45 cm diameter pipe can transport 10 megatonnes (Mt) of CO_2 per year for a cost of about \$1 per tonne of CO_2 per 100 km. By comparison, transport of small quantities in tanker trucks will cost about \$7 per tonne of CO_2 per 100 km.

Ships, like tanker trucks, transport CO_2 as a refrigerated liquid under pressure. This is because CO_2 does not exist as a liquid at atmospheric pressure. Typical conditions for the small quantities of CO_2 transported today are $-20°C$ and 20 bar. For the larger quantities that will need to be transported for BECCS, transport conditions will be closer to $-50°C$ and 7 bar. For shipping of large quantities of CO_2 distances of hundreds of kilometers or less, ship transport will be more expensive than pipelines. However, at longer distances, ship transport can become competitive. Situations that require flexibility may also favor ship transport. As an example, Norway is developing a storage facility in the North Sea that would accept CO_2 from numerous sites in several countries. It plans to use ship transport, not pipelines, because ships allow more flexibility in the development of this project.

As with most large infrastructure projects today, the permitting of CO_2 pipelines is a challenge. For example, Summit Carbon Solutions is attempting to build over 3,000

Norway is developing a storage facility in the North Sea that would accept CO_2 from numerous sites in several countries. It plans to use ship transport, not pipelines, because ships allow more flexibility in the development of this project.

kilometers of pipeline in the upper Midwest to connect thirty-two ethanol plants to a CO_2 storage site in North Dakota.[6] It has to receive permission from thousands of landowners to cross their land, and, not unexpectedly, there are holdouts. The company may need to ask state governments for eminent domain to be allowed to cross the properties of the holdouts. Despite overall political support for the project, evoking eminent domain is always controversial. Summit is confident that its $5 billion project will proceed, but it announced in October 2023 that the project will be delayed until 2026.[7]

The US Congress is considering bills that would make it easier to site a variety of infrastructure projects, including pipelines, transmission lines, and wind and solar farms. While these bills are gaining support, it is unclear whether any will become law. There are also studies looking at whether natural gas pipelines can be repurposed to carry CO_2, but they have yet to yield a definitive answer.

CO_2 pipelines have had a good safety record overall, but one high-profile accident received a bit of press. In February 2020, a pipeline that ruptured near Satartia, Mississippi, caused about forty-five people to receive medical attention.[8] The pipeline transported a mixture of CO_2 and hydrogen sulfide (H_2S), a highly toxic gas, and the H_2S caused most of the medical issues. The source of the gas was a plant where both CO_2 and H_2S were removed from natural gas, and the gas was to be injected into an oil field for enhanced oil recovery, which allows injection of the H_2S along with the CO_2. This is not the case for most CO_2 pipelines and definitely not the case for BECCS, where the CO_2 captured and transported will be very pure. CO_2 is nontoxic and nonflammable, making it a safe gas to transport. However, there is some risk as CO_2 is an asphyxiant, the implications of which are discussed in the next section.

Geological Carbon Dioxide Storage

Nature has shown that gases can be safely stored in geological formations for millions of years. The prime example is the reservoirs from which we extract natural gas for our energy needs. Most experience with the injection and storage of CO_2 to date has been in the context of carbon dioxide enhanced oil recovery (CO_2-EOR). In this process, CO_2 is injected into active oil reservoirs in order to mobilize the oil to extract a greater percentage of the original oil in place. Up until recently, there has been very limited commercial or regulatory imperative to do anything meaningful about CO_2 emissions, and thus CO_2 capture and storage projects have historically been driven by the economics of oil recovery. When oil prices are high, it makes commercial sense to squeeze a little more out of an existing oil reservoir through the use of CO_2-EOR.

In the context of CO_2 storage as part of a climate change mitigation strategy, the storage of CO_2 in deep saline formations holds the greatest potential. An example of this is the Sleipner CO_2 storage project in the Norwegian North Sea. The background here is that the Sleipner Vest gas field, discovered by Statoil (now Equinor) in 1974 and brought onstream in 1996, had a substantial CO_2 content—about 9 percent. This is often the case in natural gas recovery because natural gas often has considerable quantities of CO_2 and other acid gases. These compounds need to be removed before the natural gas can be used for energy purposes or as an industrial feedstock. This step—acid gas removal—is common across many natural gas production facilities worldwide. This CO_2 is usually vented to the atmosphere, though this is starting to change. In 1991, Norway enacted a meaningful carbon tax, and had the CO_2 from Sleipner been vented, Statoil would have had to pay a fine of one million Norwegian krone per day (approximately \$100,000 per day). Instead, the company opted to store the CO_2 in a deep saline formation known as the Utsira Formation. This project has been operational since 1996, storing about one million tonnes of CO_2 annually, and it continues to operate today.

A good geologic storage formation for CO_2 must meet four main criteria.[9] First, the target storage formation must be *porous with good permeability*. A bucket of sand meets these criteria. If you add water to the bucket, the water easily flows through the sand because it has high permeability. Pores are the space between the grains that the water can fill. Deep in the earth, sandstone formations, found around the world in sedimentary basins, exhibit similar characteristics to this bucket of sand. Second, the target storage formation must be *below 800 meters in depth*. To ensure that the CO_2 remains in a dense liquid-like phase, it needs to be stored at pressures greater than its critical pressure of 73.9 bar. The average hydrostatic pressure at 800 m depth is 80 bar, so anything deeper will safely satisfy this criterion. For CO_2 storage reservoirs, typical depths are 2 km to 3 km, with pressures of 200 bar to 300 bar and temperatures of 60°C to 100°C. This results in a range of specific gravities for the CO_2 from 0.5 to 0.8. Compare this to a CO_2 gas with a specific gravity of about 0.001 or to liquid water with a specific gravity of one. Because CO_2 is much denser than a gas, much more of it can be stored in the pore space of the formation. Being less dense than water, the CO_2 is buoyant and will want to rise up in the formation. This leads to the third criteria requiring the target storage formation to have an *impermeable caprock*. Since the CO_2 is buoyant, an impermeable caprock will keep it trapped in the target formation. Thick shale layers, consisting primarily of clay, make an excellent caprock. Fourth, the target storage formation should be *thick and continuous over large areas* in order to be able to store large volumes of CO_2.

The geological formations that meet the above criteria fall into two categories—*depleted oil and gas reservoirs* or *deep saline formations*. Oil and gas reservoirs have proved that they can hold hydrocarbons for millions of years. This gives confidence that they can store CO_2 for a long time. Because they have been producing hydrocarbons, their flow characteristics are well known. Questions do arise about whether the wells drilled into the reservoirs and the removal of the hydrocarbons have compromised their integrity. Deep saline formations are filled with salty water (i.e., saline) and are much deeper in the earth than drinking water aquifers. Because they have little or no economic value, very few wells have been drilled into them, making the data regarding their physical characteristics sparse. Ultimately, these deep saline formations will store the most CO_2 because they are widespread and have much larger volumes than oil and gas reservoirs.

Geologic storage of CO_2 is the mirror image of oil and gas production. Instead of drilling wells into the earth to extract oil and gas, wells are drilled to inject CO_2. Injection requires the CO_2 to be pressurized, typically in the range of 100 bar to 150 bar. As it moves down the well, the pressure will rise due to the weight of the CO_2 in the pipe. For injection, the pressure in the pipe must be higher than the formation pressure so that at the perforated interval at the bottom of the well, the pressure pushes the CO_2 into the formation, where it displaces water and moves into the pores of the rock, forming a plume.

Once the CO_2 enters into the formation, the laws of nature take over and determine its fate. The better we can characterize a formation in terms of its structure, dimensions, and physical properties, the better we can model how the CO_2 will move through the formation and what will happen to the plume over time. The formation will experience a pressure rise associated with the injected CO_2 displacing water. The magnitude of the pressure rise will depend on the formation properties, both in the area of the plume and far away. The pressure rise must be monitored to make sure it does not exceed the pressure that will start fracturing the rocks. In many cases, this pressure rise criterion will determine the storage capacity of a reservoir.

The number one question people ask about geologic CO_2 storage is "Will it leak?" The IPCC answers as follows: "Observations from engineered and natural analogues as well as models suggest that the fraction retained in appropriately selected and managed geological reservoirs is very likely to exceed 99% over 100 years and is likely to exceed 99% over 1000 years."[10] The IPCC defines *very likely* as 90 percent to 99 percent and *likely* as 66 percent to 90 percent. In other words, if done properly, there will be no significant leakage. So why the longwinded statement? First, not all types of projects are "appropriately managed." Second, the selection of an appropriate reservoir for geologic storage is crucial to success. Finally, since geologists think on geological timescales of millions or even billions of years, they will not guarantee that a fluid injected in the ground will never come out. There are just too many changes in earth systems on geologic timescales for that. However, if an appropriate reservoir is selected and if the project is appropriately managed, we should expect little or no leakage over human timescales of interest.

If CO_2 does leak, the impacts of elevated CO_2 levels are as follows:

> At low concentrations (less than 1% by volume), CO_2 causes no ill effects on humans, fauna or flora. In fact, CO_2 is essential for life, being a critical component in photosynthesis. Some greenhouses purposely elevate CO_2 levels in order to "fertilize" the plants. At concentrations of about 6% by volume, CO_2 can cause nausea, vomiting, diarrhea, and irritation to mucous membranes, skin lesions and sweating. At about 10% by volume, it will cause asphyxiation.[11]

Can CO_2 escaping from geologic storage reservoirs reach the elevated concentrations that cause harmful impacts? Under most cases, the answer is no. Even at high leakage rates, CO_2 leaking from the reservoir will disperse into the atmosphere and not reach harmful concentrations. However, there are two major exceptions. First, CO_2

is heavier than air, so it can gather in low-lying areas on still days. If a big enough leak occurs in a topography that will gather the CO_2, it does present a risk. The second exception is enclosed structures, such as the basement of a house. This case is analogous to the way radon can build up in basements that are not adequately ventilated. In both of these cases, monitoring can help detect the presence of CO_2 to allow any problems to be mitigated. In general, the biggest impact of CO_2 leaking from geologic storage will be the ineffectiveness of the money paid to keep that CO_2 out of the atmosphere. To date, there has been no significant leakage reported from any operating CCS projects.

There are many estimates for CO_2 storage capacity in the literature. The methodology they employ varies considerably, so the estimates are not always directly comparable. The availability and quality of data inputs are also variable. This results in large ranges given for the estimates. For example, for US storage capacity, the US Department of Energy estimates 1,877 to 14,737 gigatonnes of CO_2, while the US Geological Survey estimates 1,637 to 4,102 gigatonnes of CO_2.[12] Even at the low end of the range, these numbers are large, representing hundreds of years of US emissions.

Detailed capacity estimates have been conducted in only a few regions of the world, such as the United States and the North Sea in Europe. Jordan Kearns and his colleagues developed a methodology to extrapolate from these relatively known regions to give estimates for all regions worldwide.[13] They estimate worldwide capacity to range from 8,000 to 55,000 gigatonnes of CO_2. Even at the lower estimate, which does take into consideration pressure limitations, there are over two hundred years of storage at current worldwide emissions rates that are approaching 40 gigatonnes of CO_2 per year.

One other consideration when discussing storage capacity relates to public acceptance. Not-in-my-backyard (NIMBY) sentiments can be strong. While the world shares the benefits of CO_2 storage, the abutters assume the risks. As an analogue, look at hydraulic fracturing for oil and gas production. States like Pennsylvania and North Dakota have embraced it, while states like New York have put a moratorium on it. For the moment in Europe, offshore geologic storage projects within porous and permeable sedimentary rocks under the sea floor are operating, but onshore projects have run into resistance. Like all emerging technologies, public outreach and acceptance will be critical to its future.

Biomass-Based Carbon Dioxide Removal

There are CDR pathways that harvest biomass but do not provide energy services. In this section, we examine three of these pathways—biomass burial, bio-oil sequestration, and biochar.

First, conceptually the simplest possible approach to using biomass in a CDR pathway is *biomass burial*—that is, bury the biomass in its entirety on the basis that biomass can be sustainably harvested and then buried in an environment that significantly inhibits biomass decomposition. This approach is inspired by natural analogues, such as New Zealand kauri trees, which have been preserved in sediments for over 40,000 years. When biomass burial is proposed for at-scale use for CDR, the biomass is stored in an engineered storage chamber that can be either above or below ground. The key to this method is that the storage chamber must maintain the conditions that inhibit biomass decomposition indefinitely. These conditions often require dry, anoxic storage. The storage chamber will need to be carefully monitored to ensure the proper maintenance of storage conditions over the very long term. While there is a relatively wide range of conditions that can, in principle, inhibit biomass decomposition, a nonexhaustive list of conditions includes anoxic conditions; an absence of moisture or liquid, water, and light or UV radiation; a hypersaline environment; a pH below 5 or above 9; and a temperature below 20°C. This CDR pathway remains relatively nascent, with companies like InterEarth, Graphyte, and EnviroNZ currently performing field trials. While biomass burial is simple in concept, the challenge is to engineer the very large storage chambers that will be required cheaply enough to make the economics work.

The long-term viability of this pathway as a CO_2 removal option remains to be seen.

Earlier in this chapter, we describe the thermochemical process of gasification as a way to convert biomass to biofuels. Another thermochemical process that can be applied to biomass is *pyrolysis*,[14] where the biomass is exposed to elevated temperatures in the absence of oxygen. Here the biomass is not converted to biofuels but rather transformed into a form that is more convenient for burial. Pyrolysis breaks down the biomass into solids, liquids, and gases, the exact makeup depending on the conditions under which the pyrolysis takes place. Fast pyrolysis maximizes the production of oils and takes place at temperatures of about 500°C with residence times on the order of seconds for the produced gases. Slow pyrolysis maximizes the production of solids and takes place at temperatures up to 750°C with residence times on the order of hours. Torrefaction, which was introduced earlier in this chapter as a preprocessing step for biomass gasification, is a mild pyrolysis taking place at temperatures in the range of 200°C to 300°C and residence times of minutes to hours. In all cases, the resulting oils or solids produced for burial will contain significantly less carbon than was in the original biomass. Pyrolysis is the key technology for the bio-oil and biochar pathways described below.

Bio-oil sequestration is being developed by a company called Charm Industrial.[15] Starting in 2021, the company has been pyrolyzing waste biomass such as corn stover and turning it into bio-oil, which is injected into geologic formations. The bio-oil is easier than unprocessed biomass to "bury," but a significant amount of the carbon in the biomass is lost during pyrolysis. The produced bio-oil is a mixture of many hydrocarbons with potential toxicity that may result in environmental and permitting issues associated with subsurface injection.

The third CDR pathway that harvests biomass but does not provide energy services is *biochar*. It can deliver additional benefits as a soil amendment, unlike biomass burial and bio-oil sequestration, which deliver only carbon removals. Biochar is essentially a charcoal that is made from biomass. The practice of adding biochar to soils to improve their function and productivity goes back centuries, with interest in biochar recently increasing greatly because of its potential as a carbon removal technology. Modern biochar is made by slow pyrolysis of biomass.

Understanding the carbon removal effectiveness of biochar is complex. The sequestration potential of biochar must be understood compared to a baseline scenario without biochar, and the life-cycle greenhouse gas emissions in the production of the biochar (such as the harvesting and transport of the biomass feedstock) must be accounted for. Biochar will slowly decay once it has been applied to the soil, and it is not yet clear if biochar represents a sufficiently stable store of carbon to really count toward CDR

Biochar . . . can deliver additional benefits as a soil amendment, unlike biomass burial and bio-oil sequestration, which deliver only carbon removals.

goals. Finally, the way in which biochar interacts with the soil microbiome is vitally important. It is possible that, under certain circumstances, it could lead to increased methane emissions from the soil. Proponents see biochar becoming a significant carbon removal technology in the future, but today it is in its infancy. Many questions must be answered to allow biochar to scale as a CDR technology.

In addition to the processes outlined above, there are a number of commercial processes that produce a waste stream rich in biogenic carbon that can qualify for carbon removals. For example, any process that involves the fermentation of biomass for the production of a food or beverage, such as wine or whiskey, will produce a CO_2 by-product. It is possible to capture and store this carbon, which some organizations are doing already.[16]

Sustainable Biomass Supply

All the carbon dioxide removal pathways described in this chapter depend on a sustainable biomass supply. One of the common concerns about using forest biomass for CDR purposes is that the harvesting of biomass from a forest will reduce the carbon stock of that forest. This is often expressed as "It takes seconds to cut down a tree but decades to replace it." This is an example of a statement that is strictly factually correct but presents a highly inaccurate portrayal of the carbon dynamics of managed forests at the ecosystem scale. Sustainable biomass requires us to remove forest biomass without reducing the carbon stocks for a given piece of land. Since biomass is continually renewed (i.e., growing) in a forest, we need to make sure the rate of removal is no greater than the growth rate.

Besides the carbon accounting, sustainable biomass must be done in an environmentally responsible manner, including protecting biodiversity. As John Field and his colleagues have noted, "Economical and environmentally sustainable production at large scales requires feedstock sources with low input costs, minimal interference to existing agricultural production, and synergies with agro-ecosystem carbon storage."[17] This leads to the critical question of the amount of sustainable biomass that will be available to support biomass-based CDR.

When we talk to people in the climate community about using biomass, we sense a bit of pessimism regarding just how far it can scale. Just like oil, people see the biomass resource as finite. While this is true for both oil and biomass, if the resource is very big compared to our needs, the fact that it is finite is somewhat irrelevant. For the past hundred years, there has been fear about running out of oil, but today our oil reserves have never been greater. In talking with experts in the biomass field, we get a very positive view of biomass availability. They say that there is much more sustainable biomass available than is being used today and that the potential available resource is much larger than most nonexperts think.

To illustrate the potential for sustainable biomass, we present the results of Field and colleagues' recent white paper[18] commissioned by the Energy Futures Initiative (EFI) Foundation.[19] For the United States, they found that "with proper incentives, an estimated 0.8 billion tons to 2.2 billion tons of biomass could be produced sustainably by 2040 in total. These numbers include constraints around sustainability, including that agricultural production meets projected demand for food, animal feed, ethanol, fiber, and exports."[20] Since biomass is roughly 50 percent carbon and for each tonne of carbon we produce 3.7 tonnes of CO_2, this translates to a potential for carbon removal of about 4 billion tonnes of CO_2 annually in the US alone. While we may not deploy 100 percent of this potential, it does show that there are abundant supplies of sustainable biomass for the US, and similar results would be expected for the world.

The paper's authors estimate that 360 million dry tonnes of biomass are currently being harvested annually today in the US for bioenergy and that this number can easily double in the near term. With added incentives and using the same biomass sources, more than a billion dry tonnes can be harvested annually. If additional biomass sources are considered and if optimized technical and land management practices (including genetic improvements) are assumed, their upper estimate of 2.2 billion tonnes of biomass is obtained. The biggest source of the biomass is dedicated energy crops, but there are significant contributions from agricultural and forestry residues as well as municipal wastes.

In summary, biomass can play a significant role in helping us get to net-zero emissions. Through the carbon dioxide removal pathways described in this chapter, biomass can generate net carbon removals from the atmosphere on the billion tonnes per year scale. Some technologies, like bio-electricity with carbon capture and storage, are commercial today. What is missing is a way to monetize the carbon removals.

ENGINEERED REMOVAL PATHWAYS

The carbon removal pathways discussed in the previous two chapters rely on plants to remove carbon dioxide (CO_2) from the atmosphere via the process of photosynthesis. As discussed in chapter 2, large flows of carbon move between the atmosphere and the terrestrial biosphere, about 140 gigatonnes of carbon (GtC) per year in each direction. Modifying these flows by only a small percentage can yield large net carbon removals. Also discussed in chapter 2 was another mechanism that nature uses to remove CO_2 from the atmosphere—rock weathering. Compared to the carbon flows between the land and atmosphere, those associated with rock weathering are three orders of magnitude smaller, about 0.3 GtC per year. However, significant efforts are underway to harness the fundamental mechanism of rock weathering to remove carbon from the atmosphere and scale it to GtC per year by using engineered removal pathways.

The basic mechanism used in rock weathering is a chemical reaction between an acid (the CO_2 in the atmosphere) and a base (the alkali minerals contained in rocks). Even though the earth has an abundance of alkali minerals, they are not in a form that readily reacts with CO_2. Because of this, reaction timescales are measured in the hundreds of thousands of years. This is why the natural carbon flows due to rock weathering are so small. Therefore, one of the key challenges for engineered removal pathways is to speed up the reaction rate of CO_2 with alkali minerals.

In this chapter, we look at two major engineered removal pathways. The first pathway, direct air capture (DAC), uses alkali as a sorbent in a chemical process. This is somewhat similar to the amine scrubbing process for CO_2 capture at a bioenergy with carbon capture and storage (BECCS) power plant, as discussed in chapter 5. As with BECCS, the captured CO_2 is likely to be stored in geologic formations. In addition to the depleted oil and gas reservoirs and the deep saline formations introduced in chapter 5, in this chapter we also look at basalt formations as a storage option. The second pathway, enhanced rock weathering, uses the same alkali rocks that nature uses but attempts to speed up the process through engineered methods. Two other pathways involving the use of alkali— ocean alkalinity enhancement and direct ocean capture

(DOC)—are discussed in chapter 7 on ocean-based carbon removal pathways.

Direct Air Capture

In Hellisheidi, Iceland, a Swiss company called Climeworks has built a direct air capture (DAC) plant called Orca (see figure 8) that removes 4,000 tonnes of CO_2 per year from the atmosphere. It has been operating since September 2021, and as of early 2025, it was the largest operating DAC plant in the world. Individuals can access the Climeworks website and purchase carbon removals from Orca for the price of $1,500 per tonne of CO_2.[1] The plant is fueled by the abundant carbon-free energy available in Iceland as geothermal energy and hydropower, and the captured CO_2 is injected into the subsurface, where it reacts with the volcanic rocks (basalts) to form carbonate rocks (see the section below on basalt storage). This plant demonstrates the feasibility of DAC, but the important question is whether it can be done economically at large scale. Despite the billions of dollars being invested in DAC today, the size of the role that it will play in the future is still uncertain.

One reason for the growing interest in DAC is that it has some advantages over using biomass as a removal pathway. First, unlike some biomass pathways, DAC does not have a permanence issue, as the CO_2 is locked away

Figure 8 Climeworks' direct air capture and storage plant, Orca, in Hellisheidi, Iceland. The photo shows two of the plant's four absorber units. Each absorber unit can remove about 1,000 tonnes of carbon dioxide per year. © Climeworks.

for millennia in geologic formations. Second, the amount of CO_2 removed from the air can be measured directly. Finally, concerns regarding environmental impacts from activities like land use changes and biomass harvesting are not a concern for DAC. However, DAC does have one major concern, and that is cost.

Given the media coverage surrounding DAC, one could be forgiven for thinking that this concept is a recent innovation, but this concept was first proposed in the late 1990s and has been under increasingly intense development

Despite the billions of dollars being invested in DAC today, the size of the role it will play in the future is still uncertain.

since then. Today, over 100 DAC processes have been proposed, almost all based on a two-step process of sorption followed by desorption. In the first step, air at ambient conditions is contacted with an alkali sorbent that chemically binds with the CO_2 in the air. In the second step, steam at elevated temperatures contacts the sorbent to strip away the CO_2 from the sorbent. The sorbent can now be reused and the released CO_2 is processed so it can be stored or utilized.

What makes DAC costly is its low concentration in ambient air, roughly 420 ppm or 0.042 percent (and rising). The methods for capturing CO_2 from the air are similar to the chemical scrubbing processes used to capture CO_2 from power plant flue gases, but they have to be adapted to the low concentration of CO_2 in the air. Power plant flue gases have CO_2 concentrations in the range of 3 percent to 15 percent, well over a hundred times greater than their concentration in the air. This makes DAC much more difficult, which can be illustrated by a simple thought experiment. To represent the flue gases from a power plant, imagine a bowl containing 4,200 marbles, with 420 red marbles representing CO_2 molecules and 3,780 blue marbles representing the rest of the molecules in the power plant flue gases. These proportions are typical of the ratio of gases in the exhaust from a coal- or biomass-fired power plant. To capture the CO_2, we must remove the red marbles. To represent air, we need a bowl with 1 million marbles, with 420 red marbles representing CO_2 molecules and 999,580 blue marbles representing the other molecules in the air. You can immediately see that finding and separating these red marbles from the vast majority of blue marbles is a much more difficult task compared to the bowl of 4,200 marbles, just as capturing CO_2 molecules from air is much more difficult than capturing CO_2 molecules from flue gases.

The dilute nature of CO_2 in air introduces two fundamental, non-negotiable physical challenges that all DAC processes must address. The first challenge is the large amount of air that must be processed. At 420 ppm CO_2, the concentration of CO_2 in ambient air at 25°C and atmospheric pressure is 0.76 grams per cubic meter (g/m^3). Put another way, 1.3 million cubic meters of air contains one tonne of CO_2 (tCO_2). DAC plants typically capture 60 percent to 75 percent of the CO_2 in the air that they process, meaning that about 2 million cubic meters of air must be processed to remove one tonne of CO_2. To put this in perspective, an ordinary home window fan rated at 2,000 cubic feet per minute (cfm) would take twenty-five days of continuous operation to move 2 million cubic meters of air. In a DAC process, moving this air and contacting it with a sorbent requires large equipment sizes, which translates into significant capital costs.

The second challenge that all DAC processes must address involves energy requirements. From thermodynamics, we can calculate the minimum work required to separate CO_2 from the air, which is 133 kilowatt hours (kWh) per tonne of CO_2. This minimum work for DAC is about three times greater than the minimum work to capture CO_2 from a power plant flue gas. But this is only part of the story because one cannot operate real processes at minimum work. For example, the capture of CO_2 from a coal-fired power plant flue gas operates at about four times minimum work. There is empirical evidence that the more dilute the feed stream, the larger the ratio of actual work to the theoretical minimum work.[2] The most efficient DAC processes today operate at about eight times minimum work.[3] Since it defeats the whole purpose of DAC to have significant carbon emissions from energy use, it is essential to use energy sources containing little or no carbon. For example, if a company powered a DAC process from a coal-fired power plant, it would actually emit more CO_2 than it removed. The bottom line is that DAC processes will require significant amounts of carbon-free energy.

While there are many variations on DAC processes, they can be divided into two categories—those that use a weak base as a sorbent and those that use a strong base. Climeworks is an example of the former, using amines, which are a weak base. Amines dissolved in water are typically used as a liquid sorbent for CO_2 capture at a power plant. However, because CO_2 is so dilute in air, liquid amines are not practical for DAC. Instead, the amines are impregnated on a substrate of porous dry solids that can form *filter packs*. Fans move air through the filter packs, where the amines chemically bind with the CO_2 to remove it from the air. Once the filter packs are saturated with CO_2, they are heated to near 100°C and put under a vacuum of about 200 millibars to release the CO_2, which is pumped away. The filter packs are then ready to restart the cycle of capturing more CO_2 from the air.

While the weak base processes use a solid sorbent, processes that use a strong base can use a liquid solvent because of its greater attraction to CO_2. But the disadvantage of this strong attraction for CO_2 is that it is much harder to break the chemical bond it forms with the CO_2. An example of a strong base process is one designed by Carbon Engineering,[4] which uses an aqueous mixture of potassium hydroxide (KOH) as a solvent. The solvent contacts the air in large absorbers, removing the CO_2 by reacting it with the KOH to form potassium carbonate (K_2CO_3). To release the CO_2 and regenerate the solvent requires three chemical reactions (where *s* is solid, *g* is gas, and *aq* is dissolved in water):

$$K_2CO_3 \text{ (aq)} + Ca(OH)_2 \text{ (s)} \rightarrow 2 \text{ KOH (aq)} + CaCO_3 \text{ (s)} \quad (1)$$

$$CaCO_3 \text{ (s)} \rightarrow CaO \text{ (s)} + CO_2 \text{ (g)} \quad (2)$$

$$CaO \text{ (s)} + H_2O \rightarrow Ca(OH)_2 \text{ (s)} \quad (3)$$

The first reaction takes place in a reactor where solid calcium hydroxide ($Ca(OH)_2$) is added to the aqueous solution of potassium carbonate, causing calcium carbonate ($CaCO_3$) to precipitate out as a solid and regenerating the potassium hydroxide to go back to the absorber. The $CaCO_3$ is then filtered, dried, and sent to a reactor called a *calciner*, where the second reaction takes place at about 900°C. Finally, the calcium oxide (CaO) is hydrated in a reactor called a *slaker* to form the calcium hydroxide needed for reaction 1.

The energy to operate the calciner is usually in the form of a fossil fuel fired directly in the calciner. This means the exhaust gas from the calciner contains the CO_2 released from the $CaCO_3$, additional CO_2 from combusting the fuel, and nitrogen from the combustion air. The CO_2 from the exhaust gas of the calciner can be captured by amine scrubbing in the same way it captures CO_2 from the exhaust gas of a power plant (see chapter 5's section on bioenergy with carbon capture and storage). Another option is to use high-purity oxygen instead of air for combustion in the calciner. Using oxygen eliminates the nitrogen from the exhaust gas, ending up with an exhaust gas with high concentrations of CO_2 that is more easily captured. There is interest in oxygen-fired calciners because they have the possibility to be less costly than standard calciners using amine scrubbing. However, oxygen-fired calciners are currently a less mature technology compared to the amine scrubbing of exhaust gases. Another alternative that is under development is a calciner that uses electricity rather than fuels, leaving only the CO_2 that is released from reaction 2 exiting the calciner. In all cases, the captured CO_2 must then be stored, usually in geologic formations (see chapter 5).

Instead of using large absorbers to contact the air with a strong base, as Carbon Engineering does, some companies have taken an approach originally called *passive absorption*. This is a bit of a misnomer, as most of these companies do incorporate fans into their designs. One way to implement passive absorption is to place the sorbent on trays exposed to the ambient air. The trays are stacked in a rack, similar to the way cafeteria trays are stacked in a rack after diners have finished eating. Robots automatically load and unload the trays. The idea is to save both capital costs (by avoiding the building of large absorbers) and operating costs (by avoiding the need to move large amounts of air). However, absorbers provide an important service of efficiently contacting the air with the sorbent under controlled conditions, and it is not clear that the passive absorption approach is better or even competitive. In passive absorption, the key is to make sure enough fresh air contacts the sorbent to ensure that most of the sorbent has bonded with CO_2. Just some of the challenges include making sure a high percentage of the sorbent reacts by exposing it to air for an adequate amount of time, which is on the order of days; having enough space to hold the amount of sorbent needed to have a significant overall capture rate (very many trays are needed); maintaining the right moisture level to ensure an appreciable reaction rate; protecting the sorbent from blowing off the trays on windy days; using some fans to keep the reaction going on windless days; and figuring out exactly when the tray of sorbent has removed its quota of CO_2 and is ready to be replaced.

The question that is always asked about DAC is "What does it cost?" On the one hand, we see a price of $1,500 per tonne of CO_2 on the Climeworks website. On the other hand, we see in the literature and popular press claims that DAC can deliver carbon removals in the range of $100 to $300 per tonne of CO_2, a range we do not find credible. It must be noted that there is no single value for the cost of DAC but a range. Different processes will have different costs, and costs will also be very sensitive to the geographical location of the DAC plant. One of us (Herzog) has analyzed this question and stated, "If I had to give a range of DAC costs in 2030, my educated guess would be $600–1000 per net tCO_2 removed."[5]

Several elements drive the cost of DAC processes. First, the large amount of air to be processed leads to large equipment sizes that drive up capital costs. For the Carbon Engineering design,[6] we calculate the cross-sectional area of the absorbers to be almost 50,000 square meters to capture 1 million tonnes of CO_2 per year. This is a three-story building 10 meters high by 5 kilometers (3 miles) long. Of course, this would not be a single structure but would be divided up among multiple absorbers. However, the capital costs are large and have motivated the attempts to try the passive absorption approach discussed above.

Another major cost for DAC processes is the large amount of low- to zero-carbon energy required in the form of both heat and power (i.e., electricity). For weak base DAC processes, there are claims that heat can be supplied by cheap or free "waste heat," which may be true in some instances, but is not realistic if DAC is to be deployed at the gigatonnes of CO_2 scale because of the limited availability of waste heat. To calculate the total DAC energy requirement, one cannot simply add together the kilowatt hours of heat and power. However, there are methods for combining heat and power to come up with a single "work" requirement for DAC processes, which we estimated to be at least 1,300 kilowatt hours per tonne of CO_2[7] (this includes CO_2 compression). Let's put this number in perspective. The United States emits about 5 gigatonnes of CO_2 per year from fossil fuel combustion and industrial processes. Offsetting 20 percent of these emissions (i.e., 1 $GtCO_2$ per year) by DAC using electricity alone to satisfy the work requirement would require 1,300 terawatt hours (TWh) of electricity. In 2022, the US generated 4,231 TWh of electricity, of which 1,673 TWh were from carbon-free sources

(772 TWh nuclear, 255 TWh hydro, and 646 TWh other renewables).[8] That means for 2022, in order for DAC to remove the equivalent of 20 percent of the US's CO_2 emissions, it would require 78 percent of the carbon-free electricity the US generated. However, we need the amount of carbon-free electricity available today to grow significantly in order to displace fossil fuel use, make up for nuclear retirements, charge our electric vehicles, power our heat pumps, generate green hydrogen, and run servers for artificial intelligence applications, all in addition to powering DAC. In other words, finding the large amount of carbon-free energy required to power DAC processes is going to be a significant challenge. It also raises the question of opportunity cost in terms of how else that energy could have been used—for example, in displacing carbon-emitting fuels like coal and natural gas from the electricity grid.

For DAC processes to be economical, they need to work as close to 24/7 in all types of weather. Hardening the process to the elements adds costs. Unlike some processes that can be put in buildings to protect equipment from extreme weather, DAC processes have to be exposed to ambient air. Some suggest that since air is everywhere, DAC processes can be sited anywhere, giving them an advantage. This is an oversimplification of the complex siting process, which must consider items such as "land availability, access to low-carbon energy and other utilities like water, permitting issues, acceptable meteorological conditions, and accessibility of CO_2 storage options."[9] As DAC processes develop, there will be opportunities for cost reductions. However, as DAC grows in scale, there may be additional costs that the early small plants did not have to contend with.

DAC is attracting large investments from both the private and public sectors. For example, Occidental Petroleum was one of several private investors in Carbon Engineering. It initially acquired 25 percent of the company and in August 2023 bought the remaining 75 percent of the company for $1.2 billion. On the public side, as part of the US Bipartisan Infrastructure Law enacted in November 2021, $3.5 billion was allocated to the formation of DAC hubs in order to accelerate the demonstration and deployment of DAC technologies. In August 2023, awards of $500 million dollars each were made to establish the first two hubs: (1) South Texas DAC Hub in Kleberg County, Texas, led by 1PointFive, a subsidiary of Occidental Petroleum, and (2) Project Cypress in Calcasieu Parish, Louisiana, led by Battelle, in coordination with two DAC companies, Climeworks and Heirloom Carbon Technologies, each of which will construct a DAC plant. Heirloom's technology is based on passive absorption using a strong base. Both projects aim to remove one million tonnes of CO_2 per year and store it in deep geological formations.

Arguably, DAC enjoys substantial support from industry and government. It is attractive from a regulatory and legislative perspective owing to the conceptually simple nature of the process: You build a machine, operate it, directly remove CO_2 from the atmosphere, and permanently store it deep underground. Simple! Ultimately, costs will determine how fast DAC will grow and how big it will become.

Enhanced Rock Weathering

"Stripe, Alphabet, Shopify, and a slew of other companies plan to spend more than $57 million cumulatively to fight climate change by spreading crushed rock over farmland."[10] This news article from December 2023 reports an investment being made in Lithos Carbon, one of many companies trying to commercialize enhanced rock weathering.[11]

Natural rock weathering occurs when certain rocks are broken down by rain and wind (i.e., weathered) and then react with the CO_2 from the air to form carbonate rocks. For example, olivine, a magnesium iron silicate rock, can be weathered to form magnesium carbonate ($MgCO_3$). Basalts (volcanic rocks) are another type of rock that can react with CO_2 from the atmosphere. The problem is that the process of rock weathering in nature is slow and that most of the olivines and basalts are underground and therefore not exposed to nature's weathering process. As a result, enhanced rock weathering aims to expose more rock to the forces of weathering and to enhance the chemical reaction rates associated with weathering.

One of the simplest forms of enhanced rock weathering (see figure 9) "is to grind the rocks up, making a fine gravel or dust that reacts more easily with the air or water."[12] The ground rocks can be spread over beaches and farmers' fields and be weathered in a matter of years. Rocks like basalts may have additional benefits in increasing crop yields if spread over farmland. Certain mine tailings, a waste product from mining operations, offer a relatively inexpensive and accessible supply of alkaline rock. Ideas that are a bit more complicated—like using catalysts, both chemical and biological—are being explored to speed up the chemical reactions.

As with all other carbon removal pathways, enhanced rock weathering faces several challenges. On the positive side, permanence will not be an issue. However, monitoring, reporting, and verification (MRV) is an issue because measuring (or even estimating) how much CO_2 the rocks remove from the air is not straightforward. It is costly to mine, grind, and transport the large quantities of rock needed. Also, if fossil energy is used for these activities, they will release CO_2, thereby reducing the net amount of carbon removed. Identifying and securing large land areas in which to spread the rock may be more complex and costlier than they sound.

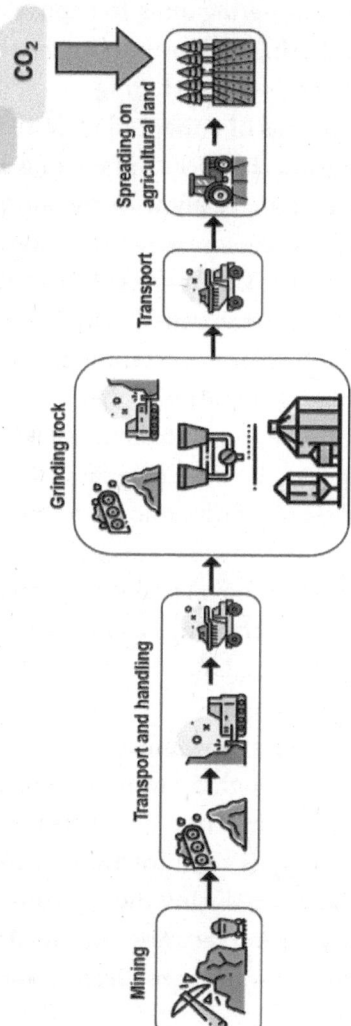

Figure 9   The five steps in enhanced rock weathering, leading to the absorption of carbon dioxide from the atmosphere.

It's also unclear how well large-scale enhanced weathering will work in practice. The chemical reactions of ground rock in natural soils or seawater are complex and hard to predict. Some studies have suggested that olivine in seawater may quickly stop combining with CO_2 in certain environments, or that using the wrong type of olivine may actually add CO_2 to the atmosphere through secondary reactions involving iron. Some byproducts of mining, grinding and applying rock may harm natural ecosystems or human health. These details need to be worked out before enhanced weathering is used at scale.[13]

While enhanced rock weathering has been under investigation since the 1990s, it is still in its infancy as far as being commercialized and deployed. In chapter 7, we explore how some of the principles involved in rock weathering are being applied to ocean-based carbon removal.

Basalt Storage

Rock weathering is a form of CO_2 mineralization, a process that turns CO_2 into carbonate minerals. Mineralization can also be used as a mechanism to store CO_2 in geologic formations, with the advantage of not having to mine, process, and transport the rock because this method brings the CO_2 to the rock. This can be thought of as *in situ* mineralization, while enhanced rock weathering is a form of *ex situ* mineralization. In the geologic storage formations discussed in chapter 5, basically sandstone formations, some mineralization does take place *in situ* but on timescales of centuries and millennia. There are other mechanisms (the most important of which involves having a secure caprock) that provide the initial containment of the injected CO_2. By contrast, basalt formations are currently being investigated because they can mineralize the CO_2 within a relatively short period of time.

As mentioned earlier in this chapter, Climeworks' DAC plant in Iceland is storing the CO_2 it removes from the air in basalt formations. It is working with Carbfix, an Icelandic company that pioneered the process. The CO_2 is injected into the volcanic basalt formations, where most is mineralized into carbonate rock within two years. The injected CO_2 must first be dissolved in water to keep it from escaping before it is mineralized. Because CO_2 has limited solubility in water, this increases the volume of injected fluids by a factor of about 25 compared to injection of pure CO_2. However, the dissolved CO_2 solution can be injected at much shallower depths than the 800 meter depth required by sandstone formations. Basalt storage is working well on the relatively small scale of thousands of tonnes per year, but it must be scaled up several orders of magnitude to become a significant storage option.

Rock weathering is a form of CO_2 mineralization, a process that turns CO_2 into carbonate minerals.

While plenty of basalt formations are distributed around the world, there is a large variation in their properties. The Icelandic basalts are well suited for CO_2 mineralization because they react relatively quickly with the CO_2. It is unclear whether basalts in other parts of the world will be as reactive or will allow easy injection of large quantities of the CO_2 solution. Basalt storage is currently being investigated in the Rift Valley in Kenya and in the Pacific Northwest in the United States. Results of these and other efforts will help answer some of the questions regarding the geographic distribution of suitable basalt storage.

OCEAN-BASED CARBON REMOVAL

As discussed in chapter 2, the ocean is by far the largest sink of "live" carbon, holding over forty-five times more carbon than the atmosphere. Therefore, it is natural to explore how to either enhance the ocean sink or remove carbon from the ocean. It is important to note that ocean-based carbon dioxide removal (CDR) methods are in their infancy and, as stated in a 2021 study from the National Academies of Sciences in the United States, "current scientific understanding of these ocean CDR approaches is insufficient to inform societal decision making."[1]

In this chapter, we explore three prominent ocean-based CDR pathways that have been proposed. Ocean alkalinity enhancement and ocean fertilization, concepts that can be traced back to the 1990s, are aimed at enhancing the ocean carbon sink. The former takes advantage of the ocean chemistry to increase the solubility of carbon dioxide (CO_2) in the ocean, while the latter induces biological activity in order to enhance the ocean's biological carbon pump. The third pathway, direct ocean capture (DOC), has emerged only in the past several years and removes carbon from the ocean, making it analogous to direct air capture (DAC), which removes carbon from the air. But first we review some fundamental principles of carbonate chemistry that help to explain the science behind ocean-based carbon removal.

An Introduction to Ocean Carbonate Chemistry

Carbon exists in the ocean in both organic form (plants and animals) and inorganic form, with about 98 percent in inorganic form. The inorganic carbon is dissolved in seawater as CO_2, bicarbonate ions (HCO_3^-), and carbonate ions ($CO_3^=$), the sum of which is called *dissolved inorganic carbon*. The dissolved inorganic carbon is partitioned between the three components such that they are in chemical equilibrium with each other as described by the following equations, with the equilibrium dependent on ocean properties including temperature, pressure, pH (acidity or basicity), and salinity.

$$CO_2 + H_2O = H_2CO_3 \qquad (4)$$

$$H_2CO_3 = H^+ + HCO_3^- \tag{5}$$

$$HCO_3^- = H^+ + CO_3^= \tag{6}$$

Roughly speaking, over 85 percent of the inorganic carbon is in the form of the bicarbonate ion, with most of the rest in the form of the carbonate ion.

This is a dynamic system where there are large flows of CO_2 between the atmosphere and ocean. However, for the purposes of this discussion, we can ignore the dynamics and consider the surface ocean's dissolved CO_2 to be in rough equilibrium with the atmospheric CO_2. As atmospheric CO_2 concentrations rise, the amount of dissolved CO_2 in the surface ocean will rise accordingly, and, as described by equations 4, 5, and 6, this will also increase the amount of bicarbonate and carbonate ions. This carbonate chemistry greatly enhances the solubility of CO_2 in the ocean, which is about 30 times greater than that of oxygen.

The ocean also contains many *cations* (ions with a positive charge)—notably, Na^+, Ca^{++}, and Mg^{++}. At high concentrations (termed *supersaturated conditions*), these cations can react with bicarbonate ions or other negatively charged ions (i.e., *anions*) to form a solid precipitate. Of particular interest is the reaction involving Ca^{++} with the bicarbonate ion because this is the mechanism by which shellfish create their shells (see equation 7). There is a common misperception that precipitating out calcite ($CaCO_3$) in seawater removes CO_2. In reality, as shown in equation 7, precipitating out calcite does just the opposite by releasing CO_2. When the shellfish dies, in most cases its shell will eventually dissolve, so the overall cycle is carbon neutral. However, the point here is that schemes to precipitate out calcite in the ocean do not remove CO_2 from the atmosphere but actually do the reverse.

$$Ca^{++} + 2\ HCO_3^- = CaCO_3(s) + CO_2 + H_2O \tag{7}$$

Ocean Alkalinity Enhancement

First proposed by Haroon Kheshgi in 1995,[2] it is well understood CO_2 can be removed from the atmosphere by adding alkalinity to the ocean. The formal term for this is *ocean alkalinity enhancement* (OAE), but since lime (CaO) is the cheapest and most widely used form of alkali in the world, this approach is often referred to as *ocean liming*. Alkali compounds are basic (the opposite of acidic), which means they have a pH greater than 7 and generate hydroxide ions (OH⁻) when dissolved in water. In the ocean, alkali compounds will react with dissolved CO_2, lowering its concentration and inducing in a net flow of CO_2 from the atmosphere to the ocean. Besides removing CO_2 from the atmosphere, adding alkalinity to the ocean will also raise the ocean pH, thereby offsetting some of the ocean acidification caused by increased concentrations of CO_2 in the oceans.

Quite a few pathways have been proposed to increase ocean alkalinity, and they all have three elements in common. First, an alkalinity source is needed. Second, the alkali must be processed, which may include steps like mining, grinding, and chemical conversion. These steps may emit CO_2, so care must be taken to minimize these emissions through use of carbon-free energy sources and/ or the capture of CO_2. Finally, the alkali must be dispersed in the ocean. In this section, we detail two prominent proposed pathways, one starting with $CaCO_3$ (commonly called calcite or limestone) and one starting with silicate rocks. In addition, we also discuss the production of alkalinity via electrochemical means.

Ocean Liming

In the previous chapter on direct air capture, we saw how sorbents composed of alkali compounds can capture CO_2 from the air. One class of these sorbents was hydroxides, and they required large absorbers, which have significant capital costs associated with them. To try to lower the capital costs, some companies are exploring "passive" absorption, which exposes the sorbent to the air without requiring big fans or expensive absorbers. But what happens if instead of exposing the sorbent to ambient air, we inject it into the ocean? The answer is that it will react with the dissolved inorganic carbon, thereby reducing its concentration and inducing a net flow of CO_2 from the atmosphere into the ocean. This effectively increases the capacity of the oceans to remove CO_2 from the atmosphere and is one form of what is referred to as *ocean liming*.

Limestone is a good choice as a source of alkali because it is inexpensive, abundant, and geographically well distributed. Processing of limestone to make lime (CaO), called *calcination*, is a well-established commercial technology, as it is a crucial step in the manufacturing of cement. This takes place in a large reactor, called a *calciner*, at temperatures of about 900°C. The reaction is shown in equation 8.

$$CaCO_3 \rightarrow CaO + CO_2 \qquad (8)$$

As can be seen in equation 8, the calcination reaction releases a significant amount of CO_2. In addition, the calciner requires significant energy to operate at high temperature and break the chemical bounds of the limestone. This is typically provided by fossil fuels, primarily coal or natural gas, which are usually directly fired in the calciner and which release additional CO_2. Since the whole goal of ocean liming is to remove CO_2 from the atmosphere, we want to minimize any CO_2 emissions to the atmosphere in making the lime. Therefore, the CO_2 produced in the calciner must be captured and stored. There are a number of approaches that can accomplish this, and they have been presented in chapter 6.

At this point, one could inject the lime into the ocean. However, it is probably better to first react the lime with water in a vessel called a *slaker* to produce calcium hydroxide or what is known as *slaked lime* $(Ca(OH)_2)$. The reaction is highly exothermic, releasing quite a bit of energy that can be recovered. If the lime was injected into the ocean without first slaking it, not only would this energy recovery opportunity be lost but there could be adverse environmental impacts by raising ocean temperatures near the injection point.

The final step is to inject the lime into the ocean. This can be done by ship or by pipe. If done within a country's territorial waters, operations would be regulated by that country's laws. However, if done in international waters, the operations would come under the London Convention (LC) and London Protocol (LP) treaties that regulate the dumping of materials at sea. These treaties set high standards for activities like ocean liming but do not prohibit it.[3]

While injection schemes using pipes may be easier and less expensive to implement, most articles on ocean liming assume injection by ships. A big reason for this is that ships allow the lime to be dispersed over larger areas of the ocean, which probably is necessary to grow these operations to the scale required to make a significant impact on climate change. Also, there are variations in the ocean with respect to temperature, pH, salinity, and currents (including upwelling and downwelling). Using ships allows the slaked lime to be injected in areas of the ocean with the most favorable conditions for it to uptake atmospheric CO_2. Injection by ships means that the governance of these operations by the LC or LP will need to be worked out, and the outcome will have a major impact on any future deployment of ocean alkalinity enhancement.[4]

The slaked lime must be dispersed well in the ocean to minimize any local environmental impacts. Of most concern is a localized spike in pH around the injection points. It is also important that the lime dissolve in the surface ocean before it sinks to greater depths. Slaked lime scores well on this point as it is made up of small particles that dissolve quickly in seawater.

Once injected and dissolved, the slaked lime will react with the CO_2 in water to form carbonate and bicarbonate ions (see equation 9). Depending on the ocean temperature, salinity, and pH at the injection point, a molecule of slaked lime will react with anywhere from 1.4 to 1.7 molecules of CO_2[5] (equation 9 assumes 1.7 molecules), which then causes a similar amount of CO_2 to be removed from the atmosphere into the ocean. Note that if the slaked lime were reacted in air, at most one molecule of CO_2 would be removed for every molecule of lime.

$$Ca(OH)_2 + 1.7\ CO_2 = Ca^{++} + 1.4\ HCO_3^- + 0.3\ CO_3^=$$
$$+ 0.3\ H_2O \hspace{4cm} (9)$$

A key question regarding ocean liming is the impact the increased alkalinity will have on the ocean environment: "The main unknown is how [ocean alkalinity enhancement] deployments would alter the biogeochemical cycling of elements on local and planetary scales and the repercussions of these alterations on marine ecosystems."[6] It should be noted that large uptakes of CO_2 by the ocean can be achieved with only small perturbations (on the order of 0.1 percent) of ocean alkalinity.[7] Most of the surface ocean is supersaturated with respect to calcite, but it does not precipitate out as $CaCO_3$ as might be expected because the precipitation is inhibited by other minerals present in the seawater. Therefore, on a global level, this small percentage increase in alkalinity is not expected to lead to $CaCO_3$ precipitation and the releasing of CO_2. Despite this, on a local level near the injection points, there is some concern that adding lime to the ocean could result in spontaneous precipitation of carbonate materials.[8] As described in the carbonate chemistry section above, this would be a major problem since it will release CO_2 to the atmosphere, which is counter to the whole purpose of ocean liming. This is another reason that the design of the system to disperse lime in the ocean is so critical. Another concern is that many plants and animals in the ocean (such as shellfish and coral) have the ability to precipitate calcite, and the exact impact on them of increased ocean alkalinity is uncertain. On the positive side, ocean liming increases ocean pH, somewhat counteracting the lowering of ocean pH (i.e., ocean acidification), a symptom of climate change caused by the increasing CO_2 concentrations in the atmosphere and the ocean.

The calcination of the limestone ($CaCO_3$) is the most expensive part of the process, so why not eliminate it and inject the limestone directly in the seawater? This is possible, but there are other issues that may justify the cost of calcining the limestone. First, limestone is not as effective as lime, reacting with only 0.4 to 0.7 molecules of CO_2 per molecule of limestone versus 1.4 to 1.7 molecules of CO_2 per molecule of lime. In addition, limestone may contain a number of impurities, such as silica, alumina, and iron, which may cause adverse environmental impacts. Finally, the limestone must be ground in an energy-intensive process to produce a fine powder—like talcum powder—to reduce the time it takes to dissolve. Even then, the dissolution may be slow because of the supersaturation in the ocean with regards to calcite, meaning that much of the limestone may sink before it dissolves.

Silicate Rocks
Silicate rocks contain calcium and/or magnesium and are the same rocks that are responsible for rock weathering in

A key question regarding ocean liming is the impact the increased alkalinity will have on the ocean environment.

the natural carbon cycle. One such rock is *olivine*, which is a magnesium iron silicate. To illustrate how it can be used to enhance ocean alkalinity, we will use a type of olivine called *forsterite* (Mg_2SiO_4) as an example.

Forsterite can be processed onshore to produce magnesium hydroxide ($Mg(OH)_2$), which can be injected into the ocean to react with dissolved CO_2 in the same manner as slaked lime,[9] with the difference being that we are replacing calcium (Ca) with magnesium (Mg). The processing of forsterite requires two major steps—mineral carbonation (see equation 10) to produce $MgCO_3$, followed by calcination—while processing limestone requires only calcination.

$$Mg_2SiO_4 + 2\,CO_2 \rightarrow 2MgCO_3 + SiO_2 \tag{10}$$

While the processing of limestone is commercially mature, the processing of forsterite requires significant development. Processing limestone is the preferred choice today, and we do not see that changing for the foreseeable future.

It has also been proposed to add forsterite directly to the ocean. In this case, the mineral carbonation (see equation 10) would occur in seawater, followed by reaction of the $MgCO_3$ with CO_2 to form carbonate and bicarbonate ions. One major challenge to this approach is the slow reaction rate for mineral carbonation in seawater. At a minimum, the forsterite must be ground very fine to try to increase reaction rates. Particles less than 10 microns would be needed for full dissolution, but producing such a fine powder would require a very substantial amount of energy. As an alternative, it has been suggested to use larger particles (~100 microns) and disperse them in coastal zones. While those particles would sink to the ocean floor, natural turbulence caused by the waves and tides would bring them back to the water column, where they would eventually dissolve over time.[10] Another concern beyond dissolving of the forsterite is that it introduces silica as well as impurities such as iron to the ocean. Implications of other minerals injected with the alkali are discussed below. As can be seen, much work is required on using silicate rocks to enhance ocean alkalinity before it can be considered a realistic option.

Electrochemical Production of Alkalinity

A battery is an example of an electrochemical device that produces electricity from the chemical reactions that take place inside that battery. Electrochemical devices called *electrolyzers* are like a battery working in reverse, meaning that electricity is used to drive chemical reactions. Electrochemistry is at the heart of the chlor-alkali electrolysis process, a large-scale industrial process used to manufacture chlorine and sodium hydroxide ($NaOH$) solution, along with a hydrogen by-product. Since sodium hydroxide can be used as the alkalinity source for ocean alkalinity enhancement, this process (or variations thereof) has been proposed to produce alkalinity via electrochemistry for injection into the ocean.[11]

A brine stream (e.g., saltwater) is required for this process. It can come from several sources, but seawater is the most likely source. While the feedstock is very inexpensive, the processing is costly, in part due to the high energy requirement in the form of electricity needed by the electrolyzers.[12] Also, a by-product of the process will be the production of acid, usually hydrochloric acid (HCl), which must be used or disposed of properly.

Dispersing the alkaline solution in the ocean must be done carefully. Since the alkali is already in solution, there is no issue with dissolution as there is for injecting alkali in solid form. However, the rate at which the alkaline solution is added to the ocean is critical. It needs to be added slowly enough to avoid precipitation of $Mg(OH)_2$ or $CaCO_3$ during injection. An experiment to inject 6,600 gallons of a sodium hydroxide solution into the waters off of Martha's Vineyard in Massachusetts is being pursued by "scientists at the highly respected Woods Hole Oceanographic Institution."[13]

Ocean Alkalinity Enhancement Issues

Many proposed ocean alkalinity enhancement schemes, like using silicate rocks, do not add pure alkali but contain other minerals (like silica, iron, and nickel) that come along with the alkali. For convenience, we call these "additional minerals." The question that needs to be answered is what impacts these additional minerals will have on the ocean environment. Minerals like nickel can be toxic in a marine environment, while iron and silica can lead to a fertilization effect by inducing biological activity. Fertilization can be positive as a potential additional pathway to remove CO_2 from the atmosphere, but it can also have adverse environmental impacts, as discussed below in the section on ocean fertilization. Additional minerals do find their way to the ocean naturally, as they may be picked up by the wind and deposited in the ocean directly or be deposited into waters that find their way to the ocean. However, the main concerns are the quantity and concentration of additional minerals that find their way to the ocean from large-scale ocean alkalinity enhancement schemes.

We can measure how much alkali we inject into the ocean, but we cannot directly measure how much CO_2 the alkali removes from the atmosphere. Questions include whether all of the alkali will react or whether some will eventually sink to the ocean bottom. If it all does react, exactly how much CO_2 will be removed per unit of alkali? The answer depends on complex ocean chemistry, which depends on temperature, salinity, and pH, which vary throughout the ocean. To be practical as a CDR pathway, credible methods must be developed to answer these and other questions related to quantifying and verifying the effectiveness of an ocean alkalinity enhancement scheme.

What will be the cost of ocean liming? The schemes described above require the extraction and processing of large amounts of minerals. The good news is that these minerals are abundant and well distributed. However, extraction and processing will have significant costs no matter what scheme is being pursued. After the mineral is processed, it must then be injected into the ocean by methods still to be determined. Whatever method is chosen, it will have significant costs associated with it. With all this uncertainty, it is hard to predict an ultimate cost. Using the commodity price of slaked lime and adding the cost to capture CO_2 in its production, we calculate a cost of $200 per net tonne of CO_2 removed. To that we need to add the cost of injecting the slaked lime into the ocean as well as other costs associated with permitting, monitoring, and other procedures. This sets a floor of about $250 per net tonne of CO_2 removed, but the actual cost could be significantly higher.

Direct Ocean Capture

If CO_2 can be removed directly from the atmosphere, why not from the ocean? After all, the ocean contains over forty-five times more carbon than the atmosphere

We can measure how much alkali we inject into the ocean, but we cannot directly measure how much CO_2 the alkali removes from the atmosphere.

contains. Furthermore, the volumetric concentration of carbon in the ocean is over a hundred times greater than that in the air. This has led to research on direct ocean capture (DOC), also referred to as direct ocean removal (DOR).

While the above numbers have spurred interest in DOC, they give just one side of the story. Other important considerations include that the carbon in the ocean is primarily in the form of carbonate and bicarbonate ions, not CO_2. Also, the carbon in the surface ocean is essentially in thermodynamic equilibrium with the atmospheric CO_2, meaning that from a thermodynamic perspective, it is equally difficult to remove carbon via DAC or DOC. While the volumetric concentration of carbon is much greater in the ocean than the atmosphere, the mass fraction is about six to seven times greater in the atmosphere. This is due to the density of seawater being about a thousand times greater than the density of air. The bottom line is that this type of top-down analysis cannot tell us if DOC is a better approach to carbon removal than DAC because we are comparing apples to oranges. We need to analyze the proposed processes for DOC to see if they are viable.

The primary pathway proposed to remove carbon from seawater is through electrochemistry. At the time of this writing, the company most advanced in developing DOC is Equatic.[14] It has two 100 kg CO_2 per day pilot facilities, one in Los Angeles and one in Singapore. In its process, an electrolyzer is used to precipitate out all the dissolved inorganic carbon in the seawater as solids in the form of $CaCO_3$ and $Mg(OH)_2$. In a second step, the carbon-free seawater is then aerated, causing it to absorb CO_2 from the air, producing more $CaCO_3$ solids, as well as $Mg(HCO_3)_2$ ions dissolved in the seawater. The processed seawater and the solid $CaCO_3$ are then returned to the ocean. Equatic states it requires 2.5 megawatt hours (MWh) of electricity per tonne of CO_2 removed, but this includes producing a hydrogen by-product with an energy content of about one MWh/tCO_2.

By its very nature, removing carbon from seawater requires a fundamental change in the seawater chemistry. To effectively analyze a DOC pathway, we must understand what happens to the seawater once it is returned to the ocean, just as we have to understand what happens to seawater in ocean alkalinity enhancement schemes. Since seawater chemistry has changed, we must understand how this changes ocean and atmospheric CO_2 flows. This is extremely challenging but necessary in order to do a complete lifecycle carbon analysis to understand the effectiveness of a given DOC process. Other important issues include environmental impacts and governance, which are described in more detail earlier in this chapter.

DOC is a very young field, with most academic papers and start-up companies appearing in just the past several years. Although it is at an early stage of development, we can still make a few observations comparing DOC to DAC.

The process pathways for DOC are less clearly defined, and it appears that their energy requirement will be at least as great as DAC, with a reasonable chance it may be greater. Working in the ocean makes the issues of environmental impacts, accounting, and governance much more complicated for DOC compared to DAC. Any numbers related to the cost of DOC right now are highly uncertain. As the field matures over the next decade or so, we will be able to better analyze its feasibility, costs, and potential as a CDR pathway.

Ocean Fertilization

"In 1988, John Martin—a respected oceanographer at the Moss Landings Laboratories—stood up at a gathering of his colleagues at the Woods Hole Oceanographic Institution, and boldly claimed *give me half a tanker of iron, and I will give you the next ice age.* With these now infamous words, Martin revitalized a theory that has become known as the *iron hypothesis.*"[15]

Martin proposed that by adding iron to the ocean, he could stimulate the ocean's biological pump (see chapter 2), thereby pulling CO_2 out of the atmosphere and cooling the planet. Some parts of the ocean are termed high nutrient, low chlorophyll (HNLC), most notably the equatorial Pacific Ocean and the Southern Ocean. *Nutrients* are compounds such as nitrogen, silicon, and phosphorous that are required for plant (e.g., phytoplankton) growth in the ocean, while *chlorophyll* refers to the phytoplankton. So why do parts of the ocean have plenty of nutrients for phytoplankton growth but a relatively low level of phytoplankton in the water? Martin proposed that there was a missing ingredient—iron.

Martin conducted two sets of experiments, dubbed the "IRONEX" experiments, in the equatorial Pacific in 1993 and 1995. The experiments did confirm that adding iron to HNLC areas of the ocean results in large phytoplankton blooms.[16] That is the basis for CDR schemes using ocean fertilization. However, using ocean fertilization to remove CO_2 from the atmosphere was controversial in the 1990s and remains so today.

If iron fertilization is used for CDR, carbon accounting will be profoundly challenging. We do know that the addition of iron to HNLC parts of the ocean can cause massive blooms of phytoplankton. This pulls CO_2 from the atmosphere, raising the question of what is the ultimate fate of the CO_2. Restating what is said in chapter 2, most of this organic material will break down in the surface layer, and about 20 percent will sink into the deep ocean, where most of it will be remineralized back to CO_2, with a very small fraction of the organic material being deposited on the ocean floor. This means that 80 percent of the CO_2 pulled from the atmosphere will find its way back to the atmosphere fairly quickly. Close to 20 percent will make it to the deep ocean, but most of that will return to the atmosphere on the order of centuries, and only a small fraction will be permanently stored on the ocean floor as organic carbon. There is significant uncertainty and variability in these numbers, making monitoring, reporting, and verification a difficult, if not impossible task.

Modeling has suggested that phytoplankton blooms in the fertilized part of the ocean deplete nutrients that would have otherwise been exported to other parts of the ocean, thereby lowering phytoplankton activity outside the fertilized area.[17] This leads to the question of how much of the increase in the biological pump in the fertilized area is truly additional. Yet another concern is the emission of other greenhouse gases—specifically, CH_4 and N_2O—from the increased biological activity, offsetting gains from CDR.

A bigger concern than the carbon accounting issues is the issue of environmental impacts. Fundamentally, the addition of iron is significantly changing the ocean ecosystem in order to achieve CDR. Some of the major issues include the following:[18]

- *Toxic effects* can come about in two ways. First, contaminants present in the large amounts of nutrients added to the ocean may lead to toxicological impacts. Second, the nutrients may stimulate naturally occurring toxins. These toxins can impair or kill zooplankton, fish, birds, or marine mammals.

• *Oxygen deficiency* is due to an increase in the amount of organic matter sinking to the deep sea. As the organic matter sinks, it increases respiration and degradation that consumes oxygen. Many marine organisms are very sensitive to low oxygen levels, which can alter or eliminate ocean biological communities.

• *Species composition and diversity* is another major concern of ocean fertilization. In the IRONEX experiments, diatoms were preferentially stimulated in relation to other species by a factor of over 85. This impacts the entire food web because diatoms are a poorer food source for some zooplankton.

While the full environmental implications of iron fertilization are unknown, our current understanding makes it hard to see how an iron fertilization commercial operation will be permitted. This point was driven home when, on May 30, 2008, at a meeting of the United Nations Convention on Biological Diversity, 191 countries agreed on a moratorium "on major ocean fertilization projects until scientists better understand the potential risks and benefits of manipulating the oceanic food chain."[19]

With these extremely large challenges that need to be addressed for iron fertilization to be considered a viable

Iron fertilization of the ocean faces large challenges that seem almost insurmountable. That is why it remains one of the most controversial approaches to CDR and why many people . . . think that iron fertilization is a nonstarter as a CDR option.

CDR option, why are people still enthused about it? The answer is money. Early estimates were that iron fertilization could be accomplished for about $2 per tonne of CO_2. This is very enticing, and in the 1990s, it did result in some now defunct companies being formed to exploit this technology. When doing full carbon accounting as described above, more recent cost estimates are in the hundreds of dollars per tonne of CO_2[20] and could easily go higher.

In summary, iron fertilization of the ocean faces large challenges that seem almost insurmountable. That is why it remains one of the most controversial approaches to CDR and why many people, including the authors of this book, think that iron fertilization is a nonstarter as a CDR option.

COMPARING CARBON REMOVAL PATHWAYS

In chapters 4 to 7, we look at many approaches for removing carbon from the atmosphere. At first glance, these approaches seem quite varied and different. However, after cutting through all the gory details and looking at the fundamentals of the different carbon removal pathways, many groupings and similarities can be identified. For this exercise, we consider the six major pathways that are presented in previous chapters—nature-based solutions, bioenergy with carbon capture and storage (BECCS), biochar, direct air capture (DAC), enhanced rock weathering, and ocean alkalinity enhancement. We have excluded direct ocean capture (DOC) and biomass burial because they are emerging pathways at very early stages of development. We also exclude ocean fertilization because of the serious questions regarding its effectiveness and environmental impacts.

The first thing to note is that there are only two fundamental ways to remove carbon from the atmosphere—either *biologically* by using photosynthesis or *chemically* by using alkali. Next, after the carbon is removed, it can be stored in either a *live* carbon reservoir or a *dead* carbon reservoir. There are two choices of live carbon reservoirs—either the *ocean* or the *terrestrial biosphere* (i.e., vegetation and soils). There are also two choices of dead carbon reservoirs—either *geological formations via injection* or *rocks via carbonation (mineralization)*.

For the biological carbon removal pathways, there are two distinct strategies. For *nature-based solutions*, trees and soils can both capture and store, with no further processing required. However, for *biochar* and *BECCS*, the biomass that captured the atmospheric carbon dioxide (CO_2) is harvested, and then additional processing is done. To qualify for net carbon removals, the harvested biomass must be contemporaneously replenished. By sustainably harvesting the biomass, a limited land resource is reused, and the amount of carbon that can be captured over time for a given plot of land is greatly increased. The sustainable harvesting of biomass is also key to the business model of many forest product industries, such as pulp and paper.

While not quite as obvious, the chemical approaches to carbon removal exhibit similar behaviors. Here, the limiting resource is not land but alkali. While alkali is not constrained in the same way as land, alkali does impose constraints because it generally is costly to produce, and its production must have a low carbon footprint. In the case of ocean alkalinity enhancement and enhanced weathering, the alkali is used once, with the captured carbon stored in place and no further processing required. For direct air capture, the alkali is used to remove CO_2 from the atmosphere, but further processing is required to release the captured CO_2 and to reuse the alkali. The released CO_2 can then be compressed, transported, and stored in geologic formations, just as is done with BECCS.

Table 3 summarizes the six major carbon removal pathways discussed above. Except for ocean alkalinity enhancement, these pathways have all been deployed at least on a pilot scale. However, they all need to be scaled up significantly to have an impact on climate.

As stated in chapter 2, there is a net emissions of about 11 gigatonnes of carbon (GtC) per year into the atmosphere, which translates to approximately 40 Gt of CO_2 every year. In addition, there are approximately an additional 10 Gt of CO_2-equivalent emissions to the atmosphere from other greenhouse gases like methane and nitrous oxide (N_2O). Therefore, for carbon removal pathways to make a significant contribution in reducing the amount of CO_2 in the atmosphere, they must work at the gigatonne scale, meaning that they should be capable of

Table 3 Comparison of six major carbon removal pathways

Removal pathway	Removal mechanism	Storage reservoir	Reuse land or alkali
Nature-based solutions	Biological	Live (land)	No
Bioenergy with carbon capture and storage (BECCS)	Biological	Dead (geological)	Yes
Biochar	Biological	Live (land)	Yes
Direct air capture (DAC)	Chemical	Dead (geological)	Yes
Enhanced rock weathering	Chemical	Dead (rocks)	No
Ocean alkalinity enhancement	Chemical	Live (ocean)	No

removing on the order of a billion tonnes of CO_2 from the atmosphere each year. While the potential is there, only nature-based solutions operate anywhere near that level today.

A key question going forward is whether these six carbon removal pathways can be scaled to the gigatonne level. To be successful, they must be able to overcome a whole set of challenges. We have clustered these challenges into five groups—permanence, accounting, cost, additionality, and permitting and governance—that are discussed below. This chapter concludes by addressing the question on whether carbon removal is a form of geoengineering.

Permanence

Because of the high expense of capturing CO_2 from the atmosphere, we want to keep it from returning to the atmosphere for a long time. But what is a long time—ten years, a hundred years, a thousand years, or longer? This is what is referred to as the issue of *permanence*, and there is no definitive answer to the question. In some ways, permanence is a misnomer for this issue. The earth is about 4.5 billion years old, and, on that time scale, nothing is really permanent. We need to look at permanence from the viewpoint of meeting our climate goals. When we emit CO_2 today, it has a climate impact. Ideally, any removed carbon should not be emitted back to the atmosphere until it will have little or no climate impact. Even with this simplified definition, quantifying permanence is difficult. If we stopped all CO_2 emissions today, it would take many millennia for the atmosphere to return to pre-industrial levels of CO_2.[1] This implies a time scale for permanence from the perspective of climate of at least a thousand years. A recent study on the durability of CDR stated "our findings suggest that a CO_2 storage period of less than 1000 years is insufficient for neutralizing remaining fossil CO_2 emissions under net zero emissions."[2]

Under certain conditions, there may be value to storage of shorter time durations, termed *temporary storage*.[3] When we emit CO_2 today, there is a cost in terms of its contribution to climate change, sometimes termed the *social cost of carbon*. If we remove CO_2 today, there will be a benefit equal to the social cost of carbon. If the captured CO_2 is then reemitted sometime in the future, the cost will be the social cost of carbon at that time. If the social cost of carbon at that time is less than it is today, then the temporary storage has some benefit. However, if the social cost of carbon is higher than it is today, then temporary storage is not worthwhile.

While the above guidelines for beneficial temporary storage are relatively straightforward, its application in the real world is extremely problematic. Calculating the social cost of carbon today is complex and controversial because many assumptions are involved and there is much uncertainty in the data.[4] Predicting the social cost of carbon in the future is even more difficult because we need to know how climate policy and technology will evolve and what their impact on CO_2 emissions will be. It is a near impossible task, which means that deciding on the value of temporary storage will always be a contentious topic.

Looking at the four storage reservoirs—land, ocean, geological, rocks—three of them have good to excellent claims of permanent storage from a climate perspective. Converting CO_2 into rocks should retain the CO_2 for millions of years, more than satisfying the criteria for permanence. Likewise, storage in the ocean as a result of alkalinity enhancement will keep the CO_2 there permanently, assuming that climate change does not change the fundamental physics and chemistry of the oceans. The International Panel for Climate Change (IPCC) has concluded that deep geological reservoirs should retain the CO_2 for over a millennium, stating: "Observations from engineered and natural analogues as well as models suggest that the fraction retained in appropriately selected and managed geological reservoirs is . . . likely to exceed 99% over 1,000 years."[5] That statement was written in 2005, and experience since that time has supported the conclusion.

The carbon removal pathway that is most challenged by the permanence criteria is the land sink encompassing soils and vegetation. The literature about storing carbon in soils and vegetation frequently cites a hundred years as the goal. That goal leads to two questions: Is the hundred-year goal achievable, and is a hundred years long enough? It seems difficult to guarantee the hundred-year goal because many of the pathways using the land sink (such as biochar) require constant maintenance. If that maintenance stops for any reason, then at least part of the captured CO_2 will likely be reemitted. Another major issue is the disturbance or destruction of the storage reservoir through events like wildfires and pestilence. The fact that climate change is increasing both the frequency and magnitude of these events is even more cause for concern.

The land sink encompassing soils and vegetation should be viewed as temporary storage. This means a tonne of carbon stored in the land sink has a lesser value than a tonne of carbon stored as a rock or injected into a geological reservoir. As described above, it is difficult to determine an accurate value for the temporary storage. Ultimately, this will be a political decision, hopefully informed by the science. Also, this issue is important when comparing costs among the various carbon removal pathways. It may be better to invest in a technology that has permanent storage compared to a technology with temporary storage, even if that permanent storage technology appears to have a significantly higher cost in terms of dollars per tonne of net CO_2 removed.

Accounting

Getting the accounting correct is essential for carbon removal to be credible. By this we mean not only accurately measuring and calculating the *gross* amount of CO_2 removed from the atmosphere but also accounting for the carbon that is emitted as part of the carbon removal project. The *net* amount of CO_2 removed is simply the gross CO_2 removed minus the CO_2 emitted. From a climate viewpoint, the net amount of CO_2 removed is what matters. As discussed elsewhere in this book, bad accounting leads some projects to be branded as greenwashing, which can tarnish even the good carbon removal projects. Three major steps are involved in accurately calculating the net amount of carbon removal to credit a project—direct measurements (measuring or estimating the gross amount of carbon removed), life-cycle emissions (calculating the amount of carbon emitted by undertaking the carbon removal project), and time dependency (accounting for the carbon removal if it is not immediate but instead takes place over an extended period of time or if some level of reversal is either likely, as with afforestation, or inevitable, as with biochar). These three steps are discussed in the following sections.

Direct Measurements

The first step for accurately calculating the net amount of carbon removal to credit a project is direct measurements. For direct air capture and BECCS, it is straightforward to obtain an accurate measurement of the gross amount of CO_2 captured, as it all flows through a pipe, where it can be measured directly by commercially available instruments. However, it is not realistic to directly measure the amount of CO_2 going into the ocean or land sinks as a result of a carbon removal project because the removal takes place over a very large area. Also, there is a large natural flow of CO_2, so we cannot really discern what flow is attributed to natural cycles versus what flow is due to carbon removal projects. For the land sink, instead of measuring the flows, it is best to measure the change in the stock of carbon.

Methodologies are available which combine direct measurements of some of the carbon stocks with models to estimate the amount of carbon removed for an entire project. Determining the carbon removal for ocean alkalinity enhancement is even more difficult, so the carbon removal must be estimated based on models. Moving away from using direct measurement to relying more on estimation and modeling methodologies results in more uncertainty in the amount of carbon removed and a greater challenge for that carbon removal pathway to overcome.

Life-Cycle Emissions

The second step for accurately calculating the net amount of carbon removal to credit a project is accounting for life-cycle emissions. This requires determining how much CO_2 and other greenhouse gases are emitted in the operation of a carbon removal project. A project's carbon footprint is determined by doing a life-cycle greenhouse gas analysis, for which there are standard methods and models. Energy use is common among most carbon removal pathways and can be a big component of life-cycle emissions. Energy use is either through direct use of fossil fuels or electricity.

It is easy and straightforward to calculate the amount of CO_2 released by burning fossil fuels. What is not as easy is to calculate the life-cycle emissions associated with the extraction, processing, and transport of the fossil fuel. As an example, natural gas is primarily methane, a potent greenhouse gas. A significant leakage of methane could potentially wipe out all benefits of the carbon removal. How much leakage occurs varies widely around the world and is not always accurately reported or easy to calculate.

To determine the carbon footprint of electricity use, one must determine the primary energy sources used in its production and the carbon footprint of those energy sources. Since the mix of primary energy sources may vary significantly over the course of a day, the carbon content of the electricity often is averaged over time. Very few places in the world have a carbon-free electricity grid today, so it is a challenge for projects to be powered solely with carbon-free electricity. One solution is to build wind or solar installations as an integral part of the project. However, since these are intermittent sources of electricity, not being able to run the carbon removal 24/7 can significantly increase the removal costs. Adding electricity storage to the projects will similarly increase costs. Another option is to buy renewable energy through a power purchase agreement and have it delivered over the grid. While the exact electrons generated from the renewable energy are not necessarily the same electrons that power the carbon removal, the arrangement works as long as the amount of renewable electricity put into the grid is the same as the amount that is taken out for carbon removal. However, if the electricity must be taken out at the same time it is produced (called *synchronous use*), it suffers the same intermittency problem as that of a project-built wind or solar installation. Therefore, some projects use *asynchronous accounting* and average use over time. For example, as long as the amount of electricity put into the grid from a renewable project over, say, a month's time balances with the amount that is removed from the grid by a carbon removal project over that same month, then all electricity for the project can be considered carbon-free. Proponents argue that this method is critical to advance the technology today. Opponents argue that this is an accounting trick and does not represent the true carbon footprint of a project. In the future, as power grids around the world decarbonize, this issue will hopefully become moot. But over the next couple of decades, it is a critical issue for all low-carbon technologies (not just carbon removal) that use electricity from the grid.

Carbon removal via biological processes have additional life-cycle considerations beyond the direct use of energy. The use of nitrogen fertilizers emits N_2O, a potent greenhouse gas. Land use change can result in direct emissions of CO_2 from the land as well as changes in the albedo, which impacts global temperature. Finally, emissions from indirect land use change can be critical but difficult to quantify. These issues are discussed in more detail in chapter 4.

Time Dependency

The third step for accurately calculating the net amount of carbon removal to credit a project is accounting for time dependency. When carbon is removed via direct air capture, the removal is immediate. This is not necessarily so with the other carbon removal pathways. When a tree is planted, its carbon removal takes place over decades, with the rate of carbon removal slowing as the tree matures. Likewise, increasing soil carbon, enhanced rock weathering, and ocean alkalinity enhancement are also time dependent, with different time scales for each of these pathways. In the case of BECCS, if biomass is harvested sustainably—meaning that biomass removals are balanced by biomass growth—then there are no time-dependency issues. However, if the rate of biomass removal is greater than the rate of biomass growth, then there is a time dependency that must be taken into account.

So how should the time dependency be calculated? Although carbon reductions can be credited year by year as they occur, project developers probably want their carbon credits as soon as possible. This is a little like lottery winners who take home much less money than the stated jackpot when they take a lump-sum payoff versus annual payments. The issue is similar to that of temporary storage, discussed above in the permanency section. For temporary storage, we are interested in the difference in the impact of emitting CO_2 today versus emitting CO_2

sometime in the future. Here we are interested in the value of removing CO_2 sometime in the future versus removing CO_2 today. Just as we concluded with permanency, doing the accounting will be somewhat subjective, and politics will almost certainly play a role.

Accounting Summary

If a company claims it has removed a net amount of CO_2 from the atmosphere, this claim must be backed up with credible and transparent data and methodologies. Each carbon removal pathway presents its own accounting challenges, with varying degrees of difficulties. If a removal pathway has a more credible accounting protocol than a competing pathway has, it may be favored even if it is more expensive.

The three key components of accounting are determining how much carbon is removed from the atmosphere (direct measurements), performing a life-cycle analysis of greenhouse gas emissions generated (life-cycle emissions), and accounting for nonpermanent and time-dependent removals (time dependency). Beyond direct measurements, it is necessary to depend on models. A balance will need to be found between how many measurements are required and how much we trust the models. Developing protocols for time dependency, whether from carbon removal over time or reemission back to the atmosphere (i.e., nonpermanence), is going to be contentious, but it is absolutely

For carbon removal to be credible and accepted by the public in the fight against climate change, we must get the accounting correct.

necessary. For carbon removal to be credible and accepted by the public in the fight against climate change, we must get the accounting correct.

Cost

Despite the many cost numbers that have been reported for carbon removal in the scientific literature as well as the popular press, we do not know what commercial-scale carbon removal will really cost. We are still in the discovery phase, and there are no commercial-scale carbon removal industries to inform the cost estimates. Cost estimates for emerging technologies tend to significantly underestimate the cost of carbon removal at commercial scale,[6] often because many real-world considerations are either minimized or absent from early cost estimates. This includes items as varied as land costs, hardening equipment for a particular environment, environmental compliance and obtaining permits, and access to the power grid.

Nature-based solutions are generally considered the least expensive carbon removal options, with costs in the tens of dollars per tonne of CO_2. This is based on purchases of credits in carbon markets today (see chapter 9). However, these markets have been harshly criticized, with people questioning the validity of the carbon credits. Also, since permanence is a major issue for nature-based

Nature-based solutions are generally considered the least expensive set of carbon removal options, with costs in the tens of dollars per tonne of CO_2.

solutions, how it impacts the cost calculations is yet to be determined. For an equivalent climate repair value, some recent estimates indicate a higher cost for nature-based solutions than for many permanent carbon dioxide removal (CDR) pathways.[7]

On the other end of the cost spectrum are the pathways that rely on a chemical-based removal mechanism. This is because the production of the chemical (alkali) that removes the CO_2 must be manufactured in processes that are both energy- and capital-intensive. These costs are estimated to be in the hundreds of dollars per tonne of CO_2 but may be even higher.

As discussed in chapter 6, we have estimated costs for direct air capture to be in the $600 to $1,000 per tonne of CO_2 range in 2030,[8] despite many claims in the popular press that costs can be as low as $100 per tonne of CO_2. This is just one example of where the hype and the hope for a technology supersedes the realities of deployment at scale. We know that the cost of CO_2 capture from atmospheric pressure sources increases as the CO_2 concentration decreases. Therefore, we can say with high confidence that the cost of capture from exhaust gases at a concentration of about 10 percent CO_2 will be significantly less than the cost of capture from the air at a concentration of 0.04 percent CO_2, a factor of dilution of 250. There have been several large-scale carbon capture projects from exhaust gases of power plants and industrial plants. These projects, which had to face real-world conditions, all have had costs greater than $100 per net tonne of CO_2 captured. This is only one of many reasons that we think direct air capture costs will be much greater than $100 per net tonne of CO_2 removed.

As we move forward in developing and scaling carbon removal technologies, the cost component will start becoming much clearer. Just as important as the cost of carbon removal technologies is the price that people and companies will be willing to pay for carbon removal. Today, they are paying tens of dollars per tonne in voluntary markets but are also paying hundreds of dollars per tonne in direct payments to carbon removal companies. This suggests that in the future there may be a market for a carbon removal technology that costs hundreds of dollars per tonne of CO_2.

Additionality

Additionality raises the question of how much of the carbon removed by a project is truly "additional" compared to a baseline of not undertaking the project. This issue generally comes up in relation to carbon markets (see chapter 9). Basically, markets do not want to pay for carbon reductions that would have occurred without their payment. Additionality is primarily a concern for nature-based solutions. For engineered removals like BECCS, direct air capture (DAC), and enhanced rock weathering, there is no way a project would be built if it were not to remove CO_2 from the atmosphere, so the traditional meaning of additionality does not apply to them.

Pathways that require significant amounts of carbon-free energy (e.g., DAC) have a different type of additionality problem. Now and for at least the next few decades, the amount of carbon-free energy is limited. That carbon-free energy is essential to decarbonize our energy systems by replacing fossil fuel electricity generation, fueling electric vehicles, powering heat pumps, and so on. That raises the question of whether it is better to use carbon-free energy to remove carbon from the atmosphere or to prevent carbon from being emitted in the first place. At least in the near term, the answer is to prevent carbon emissions. Therefore, in most cases, diverting this carbon-free energy to carbon removal will not result in additional lowering of CO_2 in the atmosphere compared to using that carbon-free energy for decarbonization of our energy systems.

Permitting and Governance

As with any project, carbon removal projects will need permission to proceed. This means obtaining permits from a governing authority as well as gaining public acceptance for the project. Even where no explicit permission is required, there will be requirements if a company wants to be compensated for carbon removal.

The projects described in previous chapters are quite varied, and therefore their permitting and governance would be expected to be quite varied. On one extreme, ocean-based carbon removal schemes will be the most difficult to permit. The oceans are governed by international law, which was established long before using the oceans for carbon removal was ever considered. This means that international law regulations must be updated to accommodate carbon removal, a slow-moving process at best. In addition, governing bodies will be cautious about approving any technology that potentially could have serious negative impacts on the ocean environment. At a minimum, a lot more work will need to be done to ensure that schemes aiming to enhance the ocean sink are both safe and effective.

Besides the ocean pathways, most of the other types of CDR projects will occur within national borders. They will have to abide by mostly existing regulations within the various countries. While obtaining permits is routine for many projects, it is sometimes hard to predict when a project will become controversial. Stakeholders may oppose a project for a variety of reasons, including concerns about environmental impacts, environmental justice considerations, not-in-my-backyard syndrome, or an opposition to the concept of carbon removal. Even after projects are approved, many countries allow court challenges that can drag on for years.

Projects can mitigate some of the permitting risk by conducting public outreach activities that inform the stakeholders of their plans, solicit their input, and respond to their concerns. While it may sound good to tell stakeholders they are helping save the world by fighting climate change, stakeholders are often more interested in the immediate risks and rewards. Saving the polar bears may sway some stakeholders, but more are swayed by the idea that jobs and tax money will be brought into their communities.

Carbon Removal as a Form of Geoengineering

Geoengineering, sometimes called *climate engineering*, refers to activities that manipulate earth systems in order to regulate climate. So is carbon removal considered to be geoengineering? The short answer is that some features of carbon removal support the label of geoengineering, but, as is discussed below, the various carbon removal pathways exhibit differentiated geoengineering aspects.

In 2015, the Committee on Geoengineering Climate of the National Research Council (NRC), part of the National Academies of Sciences, Engineering, and Medicine

Geoengineering, sometimes called *climate engineering*, refers to activities that manipulate earth systems in order to regulate climate. So is carbon removal considered to be geoengineering?

in the United States, issued two reports—*Climate Intervention: Reflecting Sunlight to Cool Earth*[9] (commonly called *solar radiation management* or SRM) and *Climate Intervention: Carbon Dioxide Removal and Reliable Sequestration*.[10] There are two major takeaways from these reports: (1) The NRC did classify carbon removal as geoengineering, and (2) the NRC deliberately separated carbon removal from SRM, stating "The NRC Committee . . . realized that Carbon Dioxide Removal and Albedo Modification [SRM] . . . have traditionally been lumped together under the term 'geoengineering' but are sufficiently different that they deserved to be discussed in separate volumes."[11]

Carbon removal is very different from solar radiation management. Carbon removal directly addresses the cause of global warming (i.e., too much CO_2 in the atmosphere) by removing that excess CO_2. SRM does nothing to address the amount of CO_2 in the atmosphere. Rather, it attempts to counteract the increased radiative forcing caused by the buildup of CO_2 in the atmosphere by reducing the amount of direct radiative forcing from the sun. SRM is controversial because it raises many issues, including the following:

• Although SRM will lower global mean temperatures, changes on the local level (such as patterns of temperature and patterns of precipitation) may be very different from the changes caused by increased greenhouse gases. There is no guarantee that the two types of changes will cancel out each other, and it is possible that they may make matters worse for many localities.

• There are concerns that SRM can cause depletion of the ozone layer.

• The issue of governance looms large since SRM will impact the whole world. Can the two hundred or so countries on earth agree to a common plan, or will a lone wolf, either governmental or from the private sector, implement an SRM scheme?

Some people argue that any time CO_2 is emitted into the atmosphere—whether through our chimneys, our tailpipes, or our factories—we are carrying out a geoengineering experiment called *climate change*. If that is the case, then the opposite action, *carbon removal*, should also be classified as geoengineering. However, lumping all carbon removal activities under the broad term *geoengineering* does not tell the whole story and is not very useful. Geoengineering requires manipulation of earth systems, but the manipulation varies widely among the various carbon removal pathways. One extreme is enhancing the ocean sink, whether through iron fertilization or ocean alkalinity enhancement, because it requires significant manipulation of ocean properties. The other extreme is direct air capture, which targets the CO_2 in the atmosphere with little to no manipulations of earth systems. This is one reason that direct air capture is receiving more attention as a carbon removal pathway.

We feel that it is a distraction to become caught up in semantics and the use of the term *geoengineering*. The four important takeaways that can be learned from this chapter's discussion are that (1) carbon removal is very different from solar radiation management, (2) some carbon removal pathways (like iron fertilization and ocean alkalinity enhancement) require significant manipulations of earth systems, (3) some carbon removal pathways (like direct air capture) require no significant manipulations of earth systems, and (4) the remaining carbon capture pathways lie somewhere in between. Whether or not carbon removal is considered geoengineering, it is important to acknowledge that each of the various carbon removal pathways has its own set of costs, benefits, and risks on which it should be judged.

LOOKING TO THE FUTURE

The need for affordable carbon dioxide removal (CDR) at scale is unquestioned. The big question is how do we get there? Addressing this point is the focus of this chapter.

As has been discussed throughout this book, there are many pathways for removing CO_2 from the atmosphere. Owing to the commercial opportunities associated with CDR, many proponents, perhaps understandably, argue for their specific carbon removal technology or approach to be considered as the solution. In reality, a winner-takes-all paradigm is highly unlikely to come to pass, and a portfolio approach is overwhelmingly likely to be the end result as it is in many other areas of human endeavor.[1] There are no silver bullets in CDR.

We envision the choice of CDR pathways will vary substantially around the world as a function of local physical and societal constraints. Moreover, the composition of

We envision the choice of CDR pathways will vary substantially around the world as a function of local physical and societal constraints.

this portfolio will evolve over time as technology develops and new pathways become available. Indeed, this is one of the primary arguments for developing and bringing to commercial readiness today as wide as possible a portfolio of potential technologies.

Voluntary Carbon Markets

The "net-zero strategies most of the world has embraced depend not just on inchoate technologies which can pull carbon dioxide out of the atmosphere and store it away, but on the creation of a carbon economy which makes doing so worthwhile."[2] Carbon markets were one of the earliest vehicles for funding CDR development and deployment. In a voluntary carbon market, companies and individuals participate in it by choice, buying and selling carbon credits to offset their emissions, rather than being legally obligated to do so under a regulatory framework.

As with many of the other concepts discussed in this book, the concept of a carbon market or an emissions trading scheme is not particularly new. Carbon markets date back to the mid-1990s, with the 1997 Kyoto Protocol establishing the first international carbon market system. Investments initially focused on reducing emissions from deforestation and forest degradation in developing countries (REDD) and from additional forest-related activities, such as sustainable managing of forests and conserving or enhancing forest carbon stocks (REDD+). The traditional geographic focus for these initiatives was Central and South America. By the early 2000s, corporations started buying credits to compensate for some of their emissions, and the first standards and registries started to appear. For example, the Gold Standard appeared in 2003, and the American Carbon Registry and Verified Carbon Standard by 2007. By 2011, the first REDD+ credits were transacted, and a number of companies started making carbon neutrality claims, often relying heavily on carbon credits to achieve these goals. In addition to REDD+ projects, the carbon markets also generated credits from emission reduction projects, such as development of renewable energy sources.[3] In 2016, the Carbon Offsetting and Reduction Scheme for International Aviation (CORSIA) was created by the International Civil Aviation Organization. This was noteworthy insofar as it was the first global market-based measure for aviation emissions, with the objective of offsetting any growth in carbon dioxide (CO_2) emissions above 2020 levels.

Globally, the carbon markets have continued to evolve, especially since nations and corporations have started making net-zero commitments. Two types of carbon markets exist today—*compliance* and *voluntary*. Compliance carbon markets are set up to support governmental programs, while voluntary carbon markets are set up and run by nongovernmental organizations. In 2021, the voluntary market was valued at about $2 billion, representing 0.5 gigatonnes (Gt) of CO_2. By 2030, its value is expected to increase by a factor of five.[4] These markets develop standards for carbon credits, verify projects are meeting these standards, and facilitate the trading of the carbon credits. In this way, entities interested in reducing their carbon footprint through the purchase of these credits can provide the capital needed to developers to undertake projects that reduce greenhouse (GHG) emissions or remove CO_2 from the atmosphere.

As carbon markets became more established, controversies soon arose. News articles called certain carbon credits "worthless,"[5] and terms like *greenwashing* were frequently used to describe the companies buying these credits. Much of the criticism revolved around two issues —*additionality* and *monitoring, reporting, and verification* (MRV)—and involved primarily REDD+ and emission reduction projects, where it is necessary to have a counterfactual (stating what might happen absent a project) to establish a baseline for carbon emissions. Credits for the project can then be measured against this baseline. However, counterfactuals by their nature contain assumptions, many of which have been questioned by critics of these markets. For example, California's carbon offset program awarded Mass Audubon credits worth about $6 million for preserving 9,700 acres of its forests from logging for the next century.[6] Realistically, Mass Audubon was never going to log its forests as this would go against its mission as an environmental organization, so these credits did not represent real carbon savings (i.e., they were not additional). Examples like this have led some to call these credits *toxic assets*, which has resulted in more interest in generating credits from engineered solutions like direct air capture (DAC) that do not require a counterfactual.

Another important point is that the conventional *offsets approach* (as opposed to carbon removal credits) is fundamentally incompatible with a net-zero paradigm. For example, even if you buy a high-quality offset from a genuinely avoided deforestation credit and then use that offset to compensate for CO_2 emissions from your airplane trip during your vacation, the net result is that CO_2 has still been added to the atmosphere, just perhaps a little less than might have otherwise occurred. This raises questions about whether an offset or REDD+ credit is truly equivalent or fungible with a removal credit and, if not, whether removal credits should have their own separate markets.

An inhibitor for CDR in the carbon markets has been a lack of consensus on what defines a quality carbon credit. Put another way, when can a carbon credit credibly offset a carbon emission and not be called a cynical attempt at greenwashing? Partially, this quandary results from the fact that contemporary efforts to develop removal credits are building on the historic carbon credit markets, which are substantially predicated on the REDD+ framework. While ecosystem conservation and restoration are worthwhile endeavors, they are distinct from carbon removal and must not be conflated.

The emerging consensus on what defines quality is focused on additionality, permanence, and MRV. All three of these topics are discussed in some detail in chapter 8. The challenge is that there is substantial variability in how these key performance indicators can be applied to and evaluated against the different CDR pathways. For example, cost aside, the pathways for afforestation, enhanced rock weathering, and DAC are very different from the perspective of the aforementioned performance indicators. In addition, each CDR pathway has different cobenefits, such as ecosystem and biodiversity services and management of soil, air, and water quality. This brings us back to the concept of fungibility. Are all carbon removal credits equal? Furthermore, can a carbon removal credit for one tonne of CO_2 fully offset a tonne of CO_2 emitted to the atmosphere? To be truly fungible, the answer to these two questions must be yes.

While fungibility of carbon credits is a goal, carbon markets are not there yet. These are complex systems, and the idea that perfect solutions are possible is unrealistic. So the question becomes what will be good enough? The issues for CDR technologies have been discussed in detail in the previous chapters of this book, including how to quantitatively account for impermanence and time-dependent removals (e.g., nature-based solutions), how to calculate net carbon removed based on life-cycle carbon analysis (e.g., DAC, BECCS), and how to estimate removals when direct measurements cannot be made (e.g., enhanced rock weathering). Finally, it will be necessary to agree on what value can be inferred from the various cobenefits that each pathway provides. If carbon markets are going to be used to finance CDR projects in the longer term, solutions that are widely accepted (i.e., good enough) must be found.

Policy and Regulation

Despite contemporary levels of attention, the policy and regulations pertaining to carbon removal remain nascent. Broadly speaking, we can decompose the world into *compliance paradigms* and *voluntary paradigms*.

Both the European Union and the United Kingdom have legislated targets for "net zero by 2050," and both regions have emissions trading. At the time of writing, CDR is not included within either of these emission trading schemes, but it is widely expected that CDR will, in due course, be included in some fashion. In the United States, there are no federally legislated targets, and the voluntary carbon market seems likely to persist in the near to medium term. In March 2024, The US Department of Energy announced its intent to launch a voluntary carbon dioxide removal purchasing challenge: "The Challenge's innovative public-private partnership structure aims to catalyze carbon dioxide removal credit purchases and improve transparency of the carbon dioxide removal credit supply."[7] In addition, legislation in the United States has created subsidies for certain low-carbon technologies. For example, CO_2 removed by direct air capture (DAC) qualifies for a tax credit (termed a *45Q tax credit*) of up to $180 per tonne of CO_2 stored in geologic formations. Many individual states have initiated climate policies, including trading systems. As in Europe, carbon removals are not yet included in those markets.

In other parts of the world, different policy approaches are likely to emerge, and it is unlikely that CDR credit prices will harmonize. Instead, the comparative advantage of different regions will come into play in determining local carbon removal credit prices. Government regulation remains sparse as stakeholders are working to understand what "good enough" looks like. However, given the critical importance and the likely scale of these markets over time, regulation is inevitable.

It is not yet obvious if there will be a reasonable level of fungibility among all carbon removals and carbon emissions or if the market will be partitioned in some way, creating a distinction between emissions trading and removals trading. It is worth restating the purpose of CDR in a net-zero paradigm—to wholly compensate for residual carbon emissions to the atmosphere. This could simply be the final percentage points of emissions from an industrial facility, such as a cement plant with CO_2 capture technology where the level of CO_2 abatement is likely to be approximately 98 percent because eliminating the final 2 percent or so may prove economically infeasible. Thus, to achieve net-zero emissions, it will be necessary to compensate for these residual emissions with carbon removals. This then begs the question of whether all forms of carbon removals are equivalent to an emission. When you burn fossil fuels, this adds a *well-defined quantity* of CO_2 to the atmosphere at a *specific point in time*, where it persists for many *thousands of years*. Arguably, therefore, in order to compensate for these carbon emissions, carbon removals need to be equally well defined in terms of the quantity of CO_2 removed, the time that it is removed, and the length of time it will stay out of the atmosphere. In table 4, we consider several different carbon dioxide removal pathways against these three criteria (see chapter 8 for details), and some clear differences emerge. In the table, a plus sign means that it is clear that the pathway meets the criteria, a zero means that the criteria can be met but the pathway requires some advanced analytics, while a dash means that the criteria are difficult to meet.[8]

Table 4 Comparison of some carbon dioxide removal pathways on the key metrics of removal timing, measurement, and permanence

- -

Carbon dioxide removal pathway	Timing of removal	Measurement of quantity removed	Permanence of removal
Bioenergy with carbon capture and storage (BECCS)	0	+	+
Direct air capture (DAC)	+	+	+
Afforestation	0	0	–
Biochar	0	+	–
Enhanced rock weathering	–	–	+

As can be readily observed from table 4, the question of equivalence and fungibility presents substantial complexity. DAC scores the highest on these criteria, which is one reason it has been drawing so much interest recently. On the other hand, afforestation has issues with all three criteria: The removal of carbon takes place over decades, the amount of carbon removed is hard to measure directly and accurately, and, under the best of circumstances, the removal will last only on the order of a century or so, assuming fires or pests will not drastically shorten that timeframe. However, looking at costs, we get a different comparison: DAC has an estimated cost in 2030 in the range of \$600 to \$1,000 per tonne of CO_2 (see chapter 6), while afforestation costs are much, much lower, around \$5 to \$20 per tonne of CO_2. So what is the best option—an expensive high-quality credit or an inexpensive credit with questionable value? Also, while credits from afforestation projects are of lower quality compared to credits from DAC projects, it is important to recognize the broader ecosystem cobenefits of afforestation. Understanding how to quantify these cobenefits and add them to carbon benefits remains an ongoing topic of research and discussion. Going forward, methodologies must be developed to credibly address the issue of fungibility if markets are going to be a mechanism to monetize carbon removals.

In a positive development for CDR, on February 20, 2024, "European Union legislators reached a political agreement . . . on a proposal to set up the world's first registry for certified carbon dioxide removals obtained from eco-farming practices and industrial processes."[9] Their proposal addresses the fungibility challenge by setting up four categories of removals:

- Permanent carbon removals, such as Direct Air Capture and Bioenergy with Carbon Capture and Storage, which can store CO_2 for several centuries.

- Temporary carbon storage in long-lasting products such as wood-based construction, for at least 35 years.

- Temporary carbon storage from carbon farming, such as restoring forests and soil, wetland management, seagrass meadows (minimum five years).

- Soil emission reduction obtained from carbon farming, such as wetland management, no tilling and cover crop practices (minimum five years).[10]

In November, 2024, the EU Council took the next step and greenlighted "a regulation that enables the establishment of the first official EU-level certification framework for permanent carbon removal, carbon farming, and CO_2 storage in products."[11] This is a good and encouraging development in addressing the issue of fungibility so that carbon removals can be incorporated into policy and regulations in an appropriate manner.

Today funding for CDR comes primarily from public subsidies in various forms (grants, tax credits, reverse auctions, etc.), as well as some investments by venture capital firms and corporations that are particularly engaged with combating climate change. While this is welcome, it is manifestly insufficient to scale markets to where they need to be. Ultimately, in order to enable CDR activity to scale, it will be necessary to move away from public subsidies and unlock private finance. In order to do this, the relevant policy and regulation must be as clear and uncomplicated as possible. Unnecessary complexity or opacity will inevitably deter private investment.

The group CDR.fyi tracks transactions of carbon removals (excluding nature-based solutions), whether they occur through a carbon market or directly between companies. As of August 2024, it recorded almost 4,000 transactions for 11 megatonnes of CO_2 worth \$3.2 billion (an average of \$290 per tonne of CO_2). However, only 3.4 percent of the orders have been filled, with biochar projects accounting for most of the removals delivered so far.[12] According to these numbers, (1) carbon removals have a very long way to go to reach the scale of gigatonnes of CO_2 per year, and (2) at least some entities are willing to pay a high price for carbon removals, a price much higher than found in carbon markets today.

Future Outlook

While climate change mitigation is an important priority, it is not the only lens through which countries consider their priorities and actions. Some countries with abundant renewable energy resources and with dynamic and well-developed economies may be willing to contemplate phasing out fossil energy. Doing so is "just" a matter of capital expenditure. However, only a handful of countries are in this position.

Today funding for CDR comes primarily from public subsidies in various forms (grants, tax credits, reverse auctions, etc.), as well as some investments by venture capital firms and corporations that are particularly engaged with combating climate change.

If we remove those preconditions, phasing out almost all fossil energy becomes an untenable proposition from economic, national, and energy security perspectives. It has recently become apparent that energy availability and reliability are of paramount importance, that countries and their political leaders are unwilling to expose their electorates to the implications of higher energy prices, and consequently that they will likely retain fossil energy in their mix for the foreseeable future, if not in perpetuity. Therefore, we urgently need to gain an understanding of the kinds of carbon removals, credits, and offsets that are fully fungible with carbon emissions.

Our expectation is that additional, verifiable, and durable or permanent removals will not be cost effective relative to the vast majority of emission reduction efforts. Thus, CDR will prove not to be an alternative to emissions reductions but rather a supplement to be utilized as necessary to reach net-zero goals. While fossil fuels will likely retain an important role in the global energy system, it will only be for the most valuable use cases.

Many elements required for the creation and scaling of the carbon markets, either voluntary or compliance, have already been resolved in other markets around the world. For example, petroleum markets trade in barrels of oil, even though the properties of the oil like viscosity and specific gravity may vary greatly. It will be important to leverage preexisting expertise rather than try to reinvent the wheel. In this context, transdisciplinary collaboration is vital.

CDR is likely to be needed at scale in the 2040s and beyond. While this may seem to be a long time from now, in the context of developing a new industry, this is a relatively short time-frame. This means it is important to start scaling CDR pathways today. In this context, policy will be vital to provide the kinds of economic incentives required to enable these technologies to proceed from their current technology readiness levels to a commercially viable proposition. As the technology evolves, the role of the public sector would be expected to evolve in parallel and move from providing direct subsidies to a more normal regulatory function.

International collaboration is going to be vital. Because there will be substantial differences in the cost of CDR across different jurisdictions, it makes sense to enable the regions that can provide carbon removal at reduced cost relative to other regions to benefit from this advantage. Political challenges notwithstanding, we think that this is going to be essential to enabling the cost-effective scale-up of CDR to the level required to address climate change concerns.

The world of carbon removal is changing quickly, and nobody has a crystal ball to predict how CDR will evolve. It will not be easy to overcome the myriad of challenges in scaling CDR to the gigatonne level, but it is possible if we have the political will. We will need to be pragmatic and focus on what works as opposed to what is perfect or "ideologically pure." It is important to recall it is not inevitable that we will meet the objective of the UN Framework Convention on Climate Change (UNFCCC) "to achieve . . . stabilization of greenhouse gas concentrations in the atmosphere at a level that would prevent dangerous anthropogenic interference with the climate system."[13] While technology is likely to play a big role in determining the future trajectory of CDR, politics will play an even bigger role. It all comes down to what the world's population chooses to do and is willing to pay. As for the debate about the role of carbon removal versus emissions abatement, it is a case of "and," not "or."

FINAL THOUGHT

While we are making progress in decarbonizing our energy systems, we are not yet on a path to keep the global mean temperature rise under 2 degrees Celsius. In other words, we are not reducing our carbon emissions fast enough. This has stimulated increased interest in carbon removal, but carbon dioxide removal (CDR) is not a white knight coming to our rescue. CDR is an important complement to emission reduction, not a substitute. If there is one message to take away from this book, it is this:

Carbon removal is essential if we want to achieve net-zero emissions.

HOWEVER, despite a myriad of pathways, it is unclear to what level carbon removal technologies can scale to. Their contributions are likely to fall far short of the hype.

THEREFORE, the best way to remove carbon dioxide from the atmosphere is not to emit it in the first place.

ACKNOWLEDGMENTS

We would like to thank the MIT Press for giving us the opportunity to write this book and our editor Beth Clevenger for all her advice and encouragement. In addition, we would like to thank our friends and colleagues who read parts or all of this book and gave us helpful feedback: Solene Chiquier (Massachusetts Institute of Technology), Robert DeRoeck (friend), Angelo Gurgel (MIT), Sergey Paltsev (MIT), Antigoni Theocharidou (Imperial College), and Joe Wrinn (friend). Finally, we would like to thank the many sponsors of our research into carbon capture and carbon removal over the past many years, the dozens of students who have worked with us, and the hundreds of colleagues from around the world whom we have had the pleasure to work with.

ACRONYMS, UNITS, AND CHEMICAL FORMULAS

Acronyms

AFOLU	agriculture, forestry, and other land uses
BECCS	bioenergy with carbon capture and storage
CCS	carbon capture and storage
CDR	carbon dioxide removal
COP	Conference of the Parties (to the UNFCCC)
DAC	direct air capture
DOC	direct ocean capture
DOR	direct ocean removal
EOR	enhanced oil recovery
ETS	Emissions Trading System
EU	European Union
GHG	greenhouse gas
HNLC	high nutrient, low chlorophyll
IPCC	Intergovernmental Panel on Climate Change
LC	London Convention
LP	London Protocol
MRV	monitoring, reporting, and verification
NRC	National Research Council
OAE	ocean alkalinity enhancement
PSA	pressure-swing adsorption
REDD	reducing emissions from deforestation and forest degradation in developing countries
REDD+	REDD plus additional forest-related activities, such as sustainable managing of forests and conserving or enhancing forest carbon stocks
SRM	solar radiation management
UNFCCC	United Nations Framework Convention on Climate Change

| UK | United Kingdom |
| US | United States |

Units

bar	A measurement of pressure equal to 0.987 atmospheres.
Celsius (C)	A measurement of temperature equal to 1.8 degrees Fahrenheit.
gram (g)	A measurement of mass. 454 grams equal 1 pound.
joule (J)	A measurement of energy. 1,060 joules equal 1 British thermal unit.
meter (m)	A measure of length equal to 3.3 feet.
micron	One millionth of a meter.
parts per billion (ppb)	A measure of concentration.
parts per million (ppm)	A measure of concentration.
tonne (t)	A metric ton. 1 tonne equals 1,000 kilograms or 2,204.6 pounds.
watt (W)	A measure of power.
watt hour (Wh)	A measure of energy equal to 3,600 joules.

Unit Prefixes

milli (m)	10^{-3}
centi (c)	10^{-2}
kilo (k)	10^{3}
mega (M)	10^{6}
giga (G)	10^{9}
tera (T)	10^{12}
exa (E)	10^{18}

Chemical Formulas

| C | carbon |
| Ca^{++} | calcium ion |

$CaCO_3$	calcium carbonate (commonly called calcite or limestone)
$Ca(HCO_3)_2$	calcium bicarbonate
CaO	calcium oxide (commonly called lime)
$Ca(OH)_2$	calcium hydroxide (commonly called slaked lime)
CH_4	methane
CO	carbon monoxide
CO_2	carbon dioxide
$CO_3^=$	carbonate ion
H_2	hydrogen
H^+	hydrogen ion
HCl	hydrochloric acid
HCO_3^-	bicarbonate ion
H_2CO_3	carbonic acid
H_2O	water or water vapor
H_2S	hydrogen sulfide
K_2CO_3	potassium carbonate
KOH	potassium hydroxide
Mg^{++}	magnesium ion
$MgCO_3$	magnesium carbonate
$Mg(HCO_3)_2$	magnesium bicarbonate
$Mg(OH)_2$	magnesium hydroxide
Mg_2SiO_4	forsterite
Na^+	sodium ion
$NaOH$	sodium hydroxide
NH_3	ammonia
N_2O	nitrous oxide
OH^-	hydroxide ion
SiO_2	silicone dioxide

Absorption
A physical or chemical separation process in which molecules enter into the bulk phase of a solvent.

Adsorption
A physical or chemical separation process in which molecules adhere to a surface.

Albedo
The fraction of light reflected from a surface.

Amines
A class of compounds that are a derivative of ammonia (NH_3). Amines replace at least one of the ammonia hydrogens with a functional group. Monoethanolamine (MEA), a sorbent used in many carbon dioxide capture processes, replaces one hydrogen with C_2H_4OH.

Anoxic
Oxygen deficient.

Anthropogenic
Resulting from human activity.

Biomass
Organic material derived from plants.

Bioenergy with carbon capture and storage (BECCS)
The application of carbon capture to bioenergy processes, such as a biomass-fired power plant or a biomass to liquid fuels process.

Caprock
In carbon storage, an impermeable rock layer that overlies the permeable rock layer where the carbon dioxide is stored.

Carbon cycle
The interchange of carbon dioxide between the atmosphere, the terrestrial biosphere (soils and vegetation), and the ocean.

Carbon capture and storage (CCS)
"A process consisting of the separation of CO_2 from industrial and energy-related sources, transport to a storage location and long-term isolation from the atmosphere" (United Nations Intergovernmental Panel on Climate Change, *Carbon Dioxide Capture and Storage*, Cambridge University Press, 2005).

Carbon footprint
The amount of carbon dioxide emissions associated with an activity or product.

Climate change
The change in the climate due to anthropogenic greenhouse gas emissions. One major climate change is the increase in global mean temperature, sometimes referred to as *global warming*.

Direct air capture (DAC)
The removal of carbon dioxide from the air by engineered systems.

Direct ocean capture (DOC)
The removal of carbon dioxide from the ocean by engineered systems.

Direct ocean removal (DOR)
Used interchangeably with direct ocean capture.

Electrochemistry
The use of electricity to drive chemical reactions (e.g., electrolysis of water to form hydrogen and oxygen). Also, the generation of electricity from chemical reactions (e.g., batteries).

Eminent domain
The taking of private property by the state for a public good.

Enhanced oil recovery (EOR)
Techniques for extracting more crude oil from an oil reservoir; sometimes called *tertiary recovery*. Carbon dioxide EOR is one such technique, where CO_2 is injected in an oil reservoir in order to mobilize the crude oil and allow it to flow to a production well.

Enhanced rock weathering
A carbon dioxide removal pathway that uses engineered systems to speed up the natural mechanism of the weathering of rocks. Rock weathering is the reaction of carbon dioxide in the atmosphere with alkali minerals in rocks to form carbonate rocks.

Flue gas
An exhaust gas of combustion, usually vented through a flue (such as a chimney or smoke stake).

Gasification
The process that converts solid fuels like coal or biomass to a gas, termed a *synthesis gas* (syngas). The syngas can be used as a feedstock to produce chemicals or fuels.

Geoengineering
The deliberate large-scale manipulation of the earth's systems to counteract the impacts of climate change.

Greenhouse gas (GHG)
A gas in the atmosphere that traps infrared radiation coming from a planet's surface, warming the planet to a temperature above what it would be without the greenhouse effect. Carbon dioxide is one such greenhouse gas.

Integrated assessment models (IAMs)
A type of energy, economic, and environment computer model that is used to assess the physical and economic impacts of climate change.

Intergovernmental Panel on Climate Change (IPCC)
A United Nations body tasked with assessing climate change science. The groups has put out a large number of reports that are available at https://www.ipcc.ch.

Legacy emissions
Emissions that are currently in the atmosphere but were emitted in the past.

Negative emissions
The removal of carbon dioxide from the atmosphere.

pH
The measurement of the acidity or basicity of a solution on a scale of 0 to 14. Low pH solutions are acidic, high pH solutions are basic, and neutral solutions have a pH of 7.

Pyrolysis
A thermochemical process that takes place at elevated temperatures in the absence of oxygen in order to break the chemical bonds of a material.

Radiative forcing
The difference between the amount of energy entering the Earth's atmosphere from the sun and the amount of energy that is emitted back into space.

Residual emissions
Greenhouse gas emissions that exist in a net-zero world because they are expensive and/or hard to abate.

Social cost of carbon
The estimate of the damage done by emitting carbon dioxide to the atmosphere. It usually is expressed in dollars per tonne of carbon dioxide ($/tCO_2$).

Synthesis gas (syngas)
A gas consisting primarily of carbon monoxide and hydrogen that is made from the gasification of coal or biomass.

Tonne
A metric ton or 1,000 kilograms (2,204.6 pounds). It is about 10 percent larger than a short ton, which weighs 2,000 pounds.

Torrefaction
A mild (i.e., low-temperature) form of pyrolysis used to turn biomass into a charcoal-like substance.

Preface

1. "United Nations Conference on Environment and Development, Rio de Janeiro, Brazil, 3–14 June 1992," United Nations, accessed January 30, 2024, https://www.un.org/en/conferences/environment/rio1992.

2. "United Nations Framework Convention on Climate Change," United Nations,1992,9,https://unfccc.int/files/essential_background/background_pub lications_htmlpdf/application/pdf/conveng.pdf.

3. Intergovernmental Panel on Climate Change, home page, accessed January 30, 2024, https://www.ipcc.ch.

4. Amrith Ramkumar and Ed Ballard, "Carbon-Removal Industry Draws Billions to Fight Climate Change," *Wall Street Journal*, June 8, 2022, https://www.wsj.com/articles/carbon-removal-industry-draws-billions-to-fight-climate-change-11654640329.

5. "CO_2 Removal Plans for Individuals," Climeworks, accessed January 30, 2024, https://climeworks.com/subscriptions.

6. At the time of this writing, the EU emissions trading system does not permit the inclusion of carbon dioxide removal, though the EU is currently working on a regulatory framework for carbon dioxide removal.

Chapter 1

1. 1 EJ (exajoule) = 10^{18} joules and 1 Btu (British thermal unit) = 1,060 joules. To power a 100 watt lightbulb for 1 hour requires 360,000 joules or 340 Btus.

2. *bp Statistical Review of World Energy*, 71st ed. (2022), 9, https://www.bp.com/content/dam/bp/business-sites/en/global/corporate/pdfs/energy-econo mics/statistical-review/bp-stats-review-2022-full-report.pdf. Note that the exact values of energy use have some uncertainty associated with them, so values from different sources may vary slightly.

3. "Annual Carbon Dioxide (CO_2) Emissions Worldwide from 1940 to 2023," Statista, accessed January 30, 2024, https://www.statista.com/statistics/276 629/global-co2-emissions.

4. "Annual Greenhouse Gas Emissions Worldwide from 1970 to 2022," Statista, accessed January 30, 2024, https://www.statista.com/statistics/128 5502/annual-global-greenhouse-gas-emissions.

5. Pierre Friedlingstein, Matthew W. Jones, Michael Sullivan, et al., "Global Carbon Budget 2021," *Earth System Science Data* 14, no. 4 (April 2022): 1951, https://doi.org/10.5194/essd-14-1917-2022.

6. Despite the difficulty of removing methane, there is some ongoing research into this option. For example, see Prachi Patel, "It's Time to Talk about Methane Removal," *Chemical & Engineering News*, February 4, 2024, https://cen.acs.org/environment/climate-change/its-time-talk-methane-removal/102/i4.

7. Samantha Eleanor Tanzer and Andrea Ramirez, "When Are Negative Emissions Negative Emissions?," *Energy & Environmental Science* 12, no. 4, (2019): 1216, https://doi.org/10.1039/C8EE03338B.

8. Kevin Andreson and Glen Peters, "The Trouble with Negative Emissions," *Science* 354, no. 6309 (October 2016): 182–183, https://www.science.org/doi/epdf/10.1126/science.aah4567.

Chapter 2

1. John D. Sterman, "Risk Communication on Climate: Mental Models and Mass Balance," *Science* 322 (October 2008): 532–533.

2. H. E. Dunsmore, "A Geological Perspective on Global Warming and the Possibility of Carbon Dioxide Removal as Calcium Carbonate Mineral," *Energy Conversion and Management* 33, no. 5–8 (1992): 565–572.

3. The quantitative numbers about the carbon cycle, as well as discussion of the mechanisms, are from the chapter of the IPCC's *Sixth Assessment Report* referenced below. The IPCC report represents the best consensus among scientists, but there is uncertainty in the values. Therefore, carbon cycle numbers from other sources may differ. Josep G. Canadell, Pedro M. S. Monteiro, Marcos H. Costa, et al., "Global Carbon and Other Biogeochemical Cycles and Feedbacks," in IPCC, *Climate Change 2021: The Physical Science Basis. Working Group I Contribution to the Sixth Assessment Report of the Intergovernmental Panel on Climate Change*, ed. V. Masson-Delmotte, V. P. Zhai, A. Pirani, et al. (Cambridge University Press, 2021), 673–816, https://doi.org/10.1017/9781009157896.007.

4. Jorge L. Sarmiento, "Ocean Carbon Cycle," *Chemical & Engineering News* 71, no. 22 (May 31, 1993): 32 https://doi.org/10.1021/cen-v071n022.p030.

Chapter 3

1. Adapted from IPCC, "Summary for Policymakers," in *Global Warming of 1.5°C. IPCC Special Report on Imports of Global Warming of 1.5°C above Pre-industrial Levels in Context of Strengthening Responses to Climate Change, Sustainable Development, and Efforts to Eradicate Poverty* (Cambridge University Press, 2018), 14, https://doi.org/10.1017/9781009157940.001.

2. Fiona Harvey, "Carbon Dioxide Removal: The Tech That Is Polarising Climate Science," *The Guardian*, April 25, 2003, https://www.theguardian.com/environment/2023/apr/25/carbon-dioxide-removal-tech-polarising-climate-science?utm_campaign=Hot%20News&utm_medium=email&_hsmi=255850420&_hsenc=p2ANqtz--wYd6Dfa_FnzAy9uHksdNdYfBYmGC1H1tX2uENy5WE-uHC7dTlSkRsqhSe_TjDhXDIIzQnOF_w8y37xqwb7pv1nIAN-g&utm_content=255850420&utm_source=hs_email.

3. Corbin Hiar, "Oil Companies Want to Remove Carbon from the Air—Using Taxpayer Dollars," *ClimateWire*, E&E News by Politico, July 13, 2023, https://www.eenews.net/articles/oil-companies-want-to-remove-carbon-from-the-air-using-taxpayer-dollars/?utm_campaign=Hot%20News&utm_medium=email&_hsmi=266592059&_hsenc=p2ANqtz--7AFSW6QLJnzIduz1wcIteHqrGUfboBfuc3t9a2nABNyFXcEdhfiGlwVMJ5wvZDh14wvc2N94M2xWiqjGnRmArVyqvDg&utm_content=266592059&utm_source=hs_email.

4. "In Line with Climate Science, Climeworks Calls for a Clear Distinction between Emissions Reductions and Carbon Removals," *Climeworks*, April 13, 2023, accessed January 31, 2024, https://climeworks.com/news/calling-for-a-clear-distinction-between-reductions-and-removals.

5. Hiar, "Oil Companies Want to Remove Carbon."

Chapter 4

1. Michael E. Miller and Frances Vinall, "In Australia's Outback, a Controversial Cash Crop Is Booming: Carbon," *Washington Post*, February 10, 2023, https://www.washingtonpost.com/world/2023/02/10/australia-carbon-farming-climate-environment.

2. John Fialka, "The Next Offset: 'Super' Poplars That Suck Up More CO_2," *ClimateWire*, February 9, 2023, https://subscriber.politicopro.com/article/eenews/2023/02/09/the-next-offset-pitch-super-poplars-that-suck-up-more-co2-00081855?utm_campaign=Hot%20News&utm_medium=email&_hsmi=245485200&_hsenc=p2ANqtz--PBQO_8PqPOVkrED6Pg-FJvQNeCyPffrqpFSEwq_v_BX0FfwbHawFdwfbVMqaHBjm8MEh6MChEIAHphgrCN_G1QcPV1A&utm_content=245485200&utm_source=hs_email.

3. 1torg, "A Platform for the Trillion Trees Community," World Economic Forum, accessed February 1, 2024, https://www.1t.org/.

4. Intergovernmental Panel on Climate Change, *IPCC Special Report: Land Use, Land-Use Change and Forestry. Summary for Policymakers* (World Meteorological Organization and United Nations Environment Program, 2000), 6, https://www.ipcc.ch/site/assets/uploads/2018/03/srl-en-1.pdf.

5. Bill Anderegg, Jeremy Freeman, Rory Jacobson, and Margaret Tory, "Improved Forest Management, Afforestation, and Reforestation," in *CDR Primer*, ed. J. Wilcox et al. (Creative Commons, 2021), section 2.4.1, https://cdrprimer.org/read/chapter-2#sec-2-4.

6. Bonnie G. Waring, Angelo Gurgel, Alexandre C. Köberle, Sergey Paltsev, and Joeri Rogelj, "Natural Climate Solutions Must Embrace Multiple Perspectives to Ensure Synergy with Sustainable Development," *Frontiers in Climate* 5 (July 13, 2023), https://doi.org/10.3389/fclim.2023.1216175.

7. Anderegg et al., "Improved Forest Management, Afforestation, and Reforestation," section 2.4.4.

8. Keith Paustian, Pete Smith, Rory Jacobson, and Margaret Torn, "Soil Carbon Sequestration," in *CDR Primer*, ed. J. Wilcox et al. (Creative Commons, 2021), section 2.3.1, https://cdrprimer.org/read/chapter-2#sec-2-3.

9. Joseph E. Fargione, Steven Bassett, Timothy Boucher, et al., "Natural Climate Solutions for the United States," *Science Advances* 4, no. 11 (November 14, 2018): 3, https://www.science.org/doi/10.1126/sciadv.aat1869.

10. Fargione et al., "Natural Climate Solutions for the United States," 5.

11. Kate Burke, "Soil Carbon Sequestration on Farms alone Won't Absolve Our Daily Emission Sins," *The Guardian*, December 18, 2021, https://www.theguardian.com/australia-news/2021/dec/19/soil-carbon-sequestration-on-farms-alone-wont-absolve-our-daily-emission-sins.

12. Jeremy Hance, "Peatland Restoration in Temperate Nations Could Be Carbon Storage Bonanza," Mongabay, February 9, 2023, https://news.mongabay.com/2023/02/peatland-restoration-in-temperate-nations-could-be-carbon-storage-bonanza/?utm_campaign=Hot%20News&utm_medium=email&_hsmi=245485200&_hsenc=p2ANqtz-9LKZHk2mjSc82vbqaLvjml7I87NiUJIyenp3ieHi9Xn1TRml-Ycf0IwhHOJZ6KRxx-ynxq55yTD41_wsZ_1BABLoFmEg&utm_content=245485200&utm_source=hs_email.

13. Tiffany Troxler, "Coastal Blue Carbon," in *CDR Primer*, ed. J. Wilcox et al. (Creative Commons, 2021), section 2.5.1, https://cdrprimer.org/read/chapter-2#sec-2-5.

Chapter 5

1. Howard J. Herzog, *Carbon Capture* (MIT Press, 2018), 39–66.

2. Mathilde Fajardy, Jennifer Morris, Angelo Gurgel, Howard Herzog, Niall Mac Dowell, and Sergey Paltsev, "The Economics of Bioenergy with Carbon Capture and Storage (BECCS) Deployment in a 1.5°C or 2°C World," *Global Environmental Change* 68 (2021): 1, https://doi.org/10.1016/j.gloenvcha.2021.102262.

3. Navid Seifker, Xiaoming Lu, Mitch Withers, et al., "Biomass to Liquid Fuels Pathways: A Techno-Economic Environmental Evaluation," MIT Energy Initiative Report, MIT, March 2015, 119, https://energy.mit.edu/publication/biomass-to-liquid-fuels-pathways/.

4. Niall Mac Dowell, Nixon Sunny, Nigel Brandon, et al., "The Hydrogen Economy: A Pragmatic Path Forward," *Joule* 5 (October 20, 2021): 2524–2539.

5. This paragraph plus the following two have been excerpted and updated from Herzog, *Carbon Capture*, 69–70.

6. Elijah Helton, "Summit Remains Upbeat on CO_2 Pipeline," *nwestiowa.com*, August 12, 2023, https://www.nwestiowa.com/news/summit-remains-upbeat-on-co2-pipeline/article_79d7ce76-93dc-11ed-bd12-6b8c9f73aec2.html.

7. Violet George, "Summit Carbon Solutions Postpones CO_2 Pipeline until 2026," *Carbon Herald*, October 19, 2023, https://carbonherald.com/summit-carbon-solutions-postpones-co2-pipeline-until-2026/.

8. Donnelle Eller, "A Carbon Dioxide Pipeline Burst in Mississippi. Here's What Happened Next," *Des Moines Register*, September 11, 2022, https://www.desmoinesregister.com/story/money/agriculture/2022/09/11/here-min ute-details-2020-mississippi-co-2-pipeline-leak-rupture-denbury-gulf-coast /8015510001.

9. This rest of this section has been excerpted and updated from Herzog, *Carbon Capture*, 70–87.

10. International Panel on Climate Change, *Carbon Dioxide Capture and Storage*, ed. Bert Metz, Ogunlade Davidson, Heleen de Coninck, Manuela Loos, and Lee Meyer (Cambridge University Press, 2005), 14, https://www.ipcc.ch/report/carbon-dioxide-capture-and-storage.

11. Jason J. Heinrich, Howard J. Herzog, and David M. Reiner, "Environmental Assessment of Geologic Storage of CO_2," Laboratory for Energy and the Environment, Massachusetts Institute of Technology, December 2003, revised March 2004, 6, http://sequestration.mit.edu/pdf/LFEE_2003-002_RP.pdf.

12. Jordan Kearns, Gary Teletzke, Jeffrey Palmer, et al., "Developing a Consistent Database for Regional Geologic CO_2 Storage Capacity Worldwide," *Energy Procedia* 114 (July 2017): 4699.

13. Kearns et al., "Developing a Consistent Database," 4697–4709.

14. Robert Nachenius, Frederik Ronsse, Robbie Venderbosch, and Wolter Prins, "Biomass Pyrolysis: Chemical Engineering for Renewables Conversion," *Advances in Chemical Engineering* 42 (2013): 75–139, https://doi.org/10.1016/B978-0-12-386505-2.00002-X.

15. Maria Gallucci, "Inside Charm Industrial's Multimillion-Dollar Bid to Remove CO_2 with Plants," Canary Media, June 19, 2023, https://www.canary

media.com/articles/carbon-capture/inside-charm-industrials-multimillion-dol
lar-bid-to-remove-co2-with-plants?utm_campaign=Hot%20News&utm_me
dium=email&_hsmi=263194273&_hsenc=p2ANqtz-_3L8ZjOKdAWkLmDz8l
UvLCg6q-X-1H7pW_yp5aPQT5NXDeYpkVSJTuGdmEgQ4O8U_-uCSBvJQZm
HQV27VQVT335mdOuw&utm_content=263194273&utm_source=hs_email.

16. The Carbon Removers (a.k.a. Carbon Capture Scotland), accessed January 7, 2025, https://thecarbonremovers.com.

17. John Field, Keith L. Kline, Matthew Langholtz, and Nagendra Singh, "Sustainably Sourcing Biomass Feedstocks for Bioenergy with Carbon Capture and Storage in the United States," UT-Battelle, LLC, June 2023, iii, https://efifoundation.org/wp-content/uploads/sites/3/2023/06/EFI_BECCS-Taking-Root_Sustainable-Feedstocks-White-Paper.pdf.

18. Field et al., "Sustainably Sourcing Biomass Feedstocks."

19. Sam F. Sartz, Tim Steeves, Nick Britton, Erik Olson, Gabrielle Easthouse, and Olafur Oddsson Cricco, "Taking Root: A Policy Blueprint for Responsible BECCS Development in the United States," EFI Foundation, June 27, 2023, https://efifoundation.org/topics/carbon-management/taking-root-a-policy-blueprint-for-responsible-beccs-development-in-the-united-states/.

20. Field et al., "Sustainably Sourcing Biomass Feedstocks," 13.

Chapter 6

1. "CO_2 Removal Plans for Individuals," *ClimeWorks*, accessed January 24, 2025, https://climeworks.com/subscriptions.

2. Kurt Zenz House, Antonio C. Baclig, Manya Ranjan, and Howard J. Herzog, "Economic and Energetic Analysis of Capturing CO_2 from Ambient Air," *Proceedings of the National Academy of Sciences* 108, no. 51 (December 2011): 20428–20433.

3. Howard Herzog, "Direct Air Capture," in *Greenhouse Gas Removal Technologies*, ed. Mai Bui and Niall Mac Dowell (Royal Society of Chemistry, 2022), 115–137.

4. David W. Keith, Geoffrey Holmes, David St. Angelo, and Kenton Heidel, "A Process for Capturing CO_2 from the Atmosphere," *Joule* 2, no. 8 (August 2018): 1573–1594, https://doi.org/10.1016/j.joule.2018.05.006.

5. Herzog, "Direct Air Capture," 136.

6. Keith et al., "A Process for Capturing CO_2," 1573–1594.

7. Herzog, "Direct Air Capture," 122–123.

8. "Table 1.1. Net Generation by Energy Source: Total (All Sectors), 2013–November 2023," US Energy Information Administration, accessed February

2, 2024, https://www.eia.gov/electricity/monthly/epm_table_grapher.php?t=table_1_01.

9. Herzog, "Direct Air Capture," 133.

10. Justine Calma, "Can Rocks Absorb Enough CO_2 to Fight Climate Change? These Companies Think So," *The Verge*, December 7, 2023, https://www.the verge.com/2023/12/7/23990979/alphabet-stripe-shopify-lithos-climate-change -carbon-removal-enhanced-weathering?_hsmi=285738321&_hsenc=p2ANqtz -_tCR2PYs_YAYsjfd18CYJvQOIXVmmwcIqjgi243nis2eTc6guzeww5DW2Djb4 HK0FXKpbyzAEirHeq5hU0uIxB0pKFBA.

11. Concepts presented in this section are from Oliver Jagoutz with Aaron Krol, "Enhanced Rock Weathering," Climate Portal, Massachusetts Institute of Technology, November 9, 2023, https://climate.mit.edu/explainers/enhanced -rock-weathering?utm_source=MIT+Energy+Initiative&utm_campaign=0409 b8c382-EMAIL_CAMPAIGN_2023_11_15_02_57&utm_medium=email&utm _term=0_-0409b8c382-%5BLIST_EMAIL_ID%5D&mc_cid=0409b8c382&mc _eid=6ebfc0acb6.

12. Jagoutz with Krol, "Enhanced Rock Weathering,"

13. Jagoutz with Krol, "Enhanced Rock Weathering,"

Chapter 7

1. National Academies of Sciences, Engineering, and Medicine, *A Research Strategy for Ocean-Based Carbon Dioxide Removal and Sequestration* (National Academies Press, 2022), 30, https://doi.org/10.17226/26278.

2. Haroon S. Kheshgi, "Sequestering Atmospheric Carbon Dioxide by Increasing Ocean Alkalinity," *Energy* 20 no. 9 (September 1995): 915–922, https://doi .org/10.1016/0360-5442(95)00035-F.

3. Lennart T. Bach, Sophie J. Gill, Rosalind E. M. Rickaby, Sarah Gore, and Phil Renforth, "CO_2 Removal with Enhanced Weathering and Ocean Alkalinity Enhancement: Potential Risks and Co-benefits for Marine Pelagic Ecosystems," *Frontiers in Climate* 1 (October 2019): 15, https://doi.org/10.3389/fclim.2019 .00007.

4. For a fuller discussion, see Katie Lebling, Eliza Northrop, Colin McCormick, and Elizabeth Bridgwater, *Toward Responsible and Informed Ocean-Based Carbon Dioxide Removal: Research and Governance Priorities* (World Resources Institute, 2022), https://doi.org/10.46830/wrirpt.21.00090.

5. Phil Renforth and Gideon Henderson, "Assessing Ocean Alkalinity for Carbon Sequestration," *Reviews of Geophysics* 55 (June 2017): 642, https://doi.org /10.1002/2016RG000533.

6. National Academies, *A Research Strategy for Ocean-Based Carbon Dioxide Removal*, 185.

7. Renforth and Henderson, "Assessing Ocean Alkalinity," 645.

8. Matthew D. Eisaman, Sonja Geilert, Phil Renforth, et al., "Assessing the Technical Aspects of Ocean-Alkalinity-Enhancement Approaches," in *Guide to Best Practices in Ocean Alkalinity Enhancement Research*, ed. A. Oschlies, A. Stevenson, L. T. Bach, et al. (Copernicus Publications, 2023), 10, https://doi.org /10.5194/sp-2-oae2023-3-2023.

9. Renforth and Henderson, "Assessing Ocean Alkalinity," 655–657.

10. Eisaman et al., "Assessing the Technical Aspects of Ocean-Alkalinity-Enhancement Approaches," 2.

11. Eisaman et al., "Assessing the Technical Aspects of Ocean-Alkalinity-Enhancement Approaches," 4–9.

12. Renforth and Henderson, "Assessing Ocean Alkalinity," 659.

13. Erin Douglas, "EPA Weighing Controversial Geoengineering Ocean Experiment South of Martha's Vineyard," *Boston Globe*, July 21, 2024, https:// www.bostonglobe.com/2024/07/21/science/geoengineering-epa-marthas-vine yard-climate-change/.

14. Equatic, accessed February 2, 2024, https://www.equatic.tech/.

15. Andrea Catherine Ryan, "Should We Fertilize the Oceans?" (master's thesis, Massachusetts Institute of Technology, 1998), 11.

16. J. H. Martin, K. H. Coale, K. S. Johnson, et al., "Testing the Iron Hypothesis in Ecosystems of the Equatorial Pacific Ocean," *Nature* 371 (1994): 123–129; Kenneth H. Coale, Kenneth S. Johnson, Steve E. Fitzwater, et al., "A Massive Phytoplankton Bloom Induced by an Ecosystem-Scale Iron Fertilization Experiment in the Equatorial Pacific Ocean," *Nature* 383 (1996): 495–501.

17. Jorge Sarmiento and James Orr, "Three-Dimensional Simulations of the Impact of Southern Ocean Nutrient Depletion on Atmospheric CO_2 and Ocean Chemistry," *Limnology and Oceanography* 36 (1999): 1928–1950, https://doi .org/10.4319/lo.1991.36.8.1928.

18. Ryan, "Should We Fertilize the Oceans?," 118–123.

19. Jeff Tollefson, "UN Decision Puts Brakes on Ocean Fertilization," *Nature* 453, 704 (June 3, 2008): 704, https://doi.org/10.1038/453704b.

20. Phillip Williamson, Philip W. Boyd, Daniel P. Harrison, Nick Reynard, and Ali Mashavek, "Feasibility of Using Biologically-Based Processes in the Open Ocean and Coastal Seas for Atmospheric CO_2 Removal," in *Greenhouse Gas Removal Technologies*, ed. Mai Bui and Niall Mac Dowell (Royal Society of Chemistry, 2022), 298–300.

Chapter 8

1. M. Inman, "Carbon Is Forever," *Nature Climate Change* 1 (December 2008): 156–158, https://doi.org/10.1038/climate.2008.122.

2. C. Brunner, Z. Hausfather, and R. Knutti, "Durability of Carbon Dioxide Removal Is Critical for Paris Climate Goals," *Communications Earth & Environment* 5, no. 645 (2024), https://doi.org/10.1038/s43247-024-01808-7.

3. Howard J. Herzog, Ken Caldeira, and J. Reilly, "An Issue of Permanence: Assessing the Effectiveness of Temporary Carbon Storage," *Climatic Change* 59, no. 3 (August 2003): 293–310.

4. David Pearce, "The Social Cost of Carbon," in *Climate-Change Policy*, ed. Dieter Helm (Oxford University Press, 2010).

5. International Panel on Climate Change, *Carbon Dioxide Capture and Storage*, ed. Bert Metz, Ogunlade Davidson, Heleen de Coninck, Manuela Loos, and Leo Meyer (Cambridge University Press, 2005), 14, https://www.ipcc.ch/report/carbon-dioxide-capture-and-storage.

6. Edward S. Rubin, "Improving Cost Estimates for Advanced Low-Carbon Power Plants," *International Journal of Greenhouse Gas Control* 88 (September, 2019): 1–9, https://doi.org/10.1016/j.ijggc.2019.05.019.

7. Augustin Prado and Niall Mac Dowell, "The Cost of Permanent Carbon Dioxide Removal," *Joule* 7 (April 19, 2023): 700–712, https://doi.org/10.1016/j.joule.2023.03.006.

8. Howard Herzog, "Direct Air Capture," in *Greenhouse Gas Removal Technologies*, ed. Mai Bui and Niall Mac Dowell (Royal Society of Chemistry, 2022), 136.

9. National Research Council, *Climate Intervention: Reflecting Sunlight to Cool Earth* (National Academies Press, 2015), https://doi.org/10.17226/18988.

10. National Research Council, *Climate Intervention: Carbon Dioxide Removal and Reliable Sequestration* (National Academies Press, 2015), https://doi.org/10.17226/18805.

11. National Research Council, *Climate Intervention: Carbon Dioxide Removal*, ix.

Chapter 9

1. Solene Chiquier, Mathilde Fajardy, and Niall Mac Dowell, "CO_2 Removal and 1.5°C: What, When, Where, and How?," *Energy Advances*, 1, no. 8 (June 14, 2022): 524–561, https://doi.org/10.1039/D2YA00108J.

2. "Special Report Carbon Dioxide Removal," *The Economist*, November 25, 2023, 4.

3. Teresa Hartmann and Douglas Broom, "What Are Carbon Credits and How Can They Help Fight Climate Change?," World Economic Forum, November 12,

2020, https://www.weforum.org/agenda/2020/11/carbon-credits-what-how -fight-climate-change/.

4. Anders Porsborg-Smith, Jesper Nielsen, Bayo Owolabi, and Carl Clayton, "The Voluntary Carbon Market Is Thriving," Boston Consulting Group, January 19, 2023, https://www.bcg.com/publications/2023/why-the-voluntary-carbon -market-is-thriving.

5. Patrick Greenfield, "Revealed: More Than 90% of Rainforest Carbon Offsets by Biggest Certifier Are Worthless, Analysis Shows," *The Guardian*, January 18, 2023, https://www.theguardian.com/environment/2023/jan/18/revealed -forest-carbon-offsets-biggest-provider-worthless-verra-aoe.

6. Lisa Song and James Temple, "A Nonprofit Promised to Preserve Wildlife. Then It Made Millions Claiming It Could Cut Down Trees," *MIT Technology Review*, May 10, 2021, https://www.technologyreview.com/2021/05/10/102 4751/carbon-credits-massachusetts-audubon-california-logging-co2-emissions -increase.

7. "U.S. Department of Energy Announces Intent to Launch Voluntary Carbon Dioxide Removal Purchasing Challenge," Fossil Energy and Carbon Management, US Department of Energy, March 14, 2024, https://content.gov delivery.com/accounts/USDOEOFE/bulletins/390679b.

8. For bioenergy with carbon capture and storage (BECCS) and biochar, we assume that they incorporate a sustainable biomass source. As a result, determining the quantity of carbon removed via measurements and life-cycle analysis is straightforward. However, the timing of removal depends on how their sustainable biomass supply chain is managed, so that is why they are rated zero in this category.

9. Frédéric Simon, "The EU's New Carbon Removal Certification Scheme in Detail," Euractiv, February 20, 2024, https://www.euractiv.com/section /climate-environment/news/eu-reaches-deal-on-worlds-first-carbon-removal -certification-scheme/?utm_source=IEAGHG+Newsletter&utm_campaign=8e80 9116d0-EMAIL_WeeklyNews_21_02_2024&utm_medium=email&utm_term =0_fc3cfeb016-8e809116d0-60566753.

10. Simon, "The EU's New Carbon Removal Certification Scheme in Detail."

11. Sasha Ranevska, "EU Approves First Official Certification Framework for Carbon Removals," *Carbon Herald*, November 20, 2024, https://carbonherald .com/eu-approves-first-official-certification-framework-for-carbon-removals.

12. CDR.fyi, accessed August 27, 2024, https://www.cdr.fyi.

13. "United Nations Framework Convention on Climate Change," United Nations, 1992, 9, https://unfccc.int/files/essential_background/background _publications_htmlpdf/application/pdf/conveng.pdf.

FURTHER READING

Bui, Mai, and Niall Mac Dowell, eds. *Greenhouse Gas Removal Technologies*. Royal Society of Chemistry, 2022. https://doi.org/10.1039/9781839165245.

Fuss, Sabine, et al. "Negative Emissions—Part 2: Costs, Potentials and Side Effects." *Environmental Research Letters* 13, no. 6 (May 22, 2018). https://doi.org/10.1088/1748-9326/aabf9f.

Minx, Jan C., et al. "Negative Emissions—Part 1: Research Landscape and Synthesis." *Environmental Research Letters* 13, no. 6 (May 22, 2018). https://doi.org/10.1088/1748-9326/aabf9b.

National Academies of Sciences, Engineering, and Medicine. *Negative Emissions Technologies and Reliable Sequestration: A Research Agenda*. National Academies Press, 2019. https://doi.org/10.17226/25259.

National Academies of Sciences, Engineering, and Medicine. *A Research Strategy for Ocean-Based Carbon Dioxide Removal and Sequestration*. National Academies Press, 2022. https://doi.org/10.17226/26278.

Royal Society and Royal Academy of Engineering. *Greenhouse Gas Removal*. 2018. https://royalsociety.org/news-resources/projects/greenhouse-gas-removal.

Wilcox, J., B. Kolosz, and J. Freeman, eds. *Carbon Dioxide Removal Primer*. Creative Commons, 2021. https://cdrprimer.org.

Zakkour, Paul, and Greg Cook. "Measurement, Reporting and Verification and Accounting for Carbon Dioxide Removal in the Context of Both Project-based Approaches and National Greenhouse Gas Inventories." IEAGHG Technical Report 2024-09, October 2024. https://ieaghg.org/publications/measurement-reporting-and-verification-and-accounting-for-carbon-dioxide-removal.

INDEX

Note: Page numbers in *italics* indicate tables or figures.

HOWARD J. HERZOG is a senior research engineer in the MIT Energy Initiative. He was a coordinating lead author for the Intergovernmental Panel on Climate Change's special report on *Carbon Dioxide Capture and Storage* (2005) and a US delegate to the Carbon Sequestration Leadership Forum's technical group (2003–2007). He received the 2010 Greenman Award from the International Energy Agency Greenhouse Gas R&D Program "in recognition of contributions made to the development of greenhouse gas control technologies." His book *Carbon Capture* was published by the MIT Press in 2018.

NIALL MAC DOWELL is a professor in energy systems engineering at Imperial College London. He is a chartered engineer and a fellow of both the Institution of Chemical Engineers and the Royal Society of Chemistry. His research is focused on understanding the transition to a low carbon economy, and he has published more than two hundred peer-reviewed scientific papers, technical reports, and books in this context. He has worked with private-sector organizations in the energy industry and financial sector and with the UK government as an expert policy adviser on carbon dioxide capture and storage as well as carbon dioxide removal.

Publisher contact:
The MIT Press
Massachusetts Institute of Technology
77 Massachusetts Avenue, Cambridge, MA 02139
mitpress.mit.edu

EU Authorised Representative:
Easy Access System Europe, Mustamäe tee 50,
10621 Tallinn, Estonia
gpsr.requests@easproject.com

Printed by Integrated Books International,
United States of America